Uncertainty Quantification
of Stochastic Defects
in Materials

Emerging Materials and Technologies

Series Editor:
Boris I. Kharissov

Biomaterials and Materials for Medicine: Innovations in Research, Devices, and Applications
Jingan Li

Advanced Materials and Technologies for Wastewater Treatment
Sreedevi Upadhyayula and Amita Chaudhary

Green Tribology: Emerging Technologies and Applications
T.V.V.L.N. Rao, Salmiah Binti Kasolang, Xie Guoxin, Jitendra Kumar Katiyar, and Ahmad Majdi Abdul Rani

Biotribology: Emerging Technologies and Applications
T.V.V.L.N. Rao, Salmiah Binti Kasolang, Xie Guoxin, Jitendra Kumar Katiyar, and Ahmad Majdi Abdul Rani

Bioengineering and Biomaterials in Ventricular Assist Devices
Eduardo Guy Perpétuo Bock

Semiconducting Black Phosphorus: From 2D Nanomaterial to Emerging 3D Architecture
Han Zhang, Nasir Mahmood Abbasi, and Bing Wang

Biomass for Bioenergy and Biomaterials
Nidhi Adlakha, Rakesh Bhatnagar, and Syed Shams Yazdani

Energy Storage and Conversion Devices: Supercapacitors, Batteries, and Hydroelectric Cell
Anurag Gaur, A.L. Sharma, and Anil Arya

Nanomaterials for Water Treatment and Remediation
Srabanti Ghosh, Aziz Habibi-Yangjeh, Swati Sharma, and Ashok Kumar Nadda

2D Materials for Surface Plasmon Resonance-Based Sensors
Sanjeev Kumar Raghuwanshi, Santosh Kumar, and Yadvendra Singh

Functional Nanomaterials for Regenerative Tissue Medicines
Mariappan Rajan

Uncertainty Quantification of Stochastic Defects in Materials
Liu Chu

For more information about this series, please visit: https://www.routledge.com/Emerging-Materials-and-Technologies/book-series/CRCEMT

Uncertainty Quantification of Stochastic Defects in Materials

Liu Chu

CRC Press
Taylor & Francis Group
Boca Raton London New York

CRC Press is an imprint of the
Taylor & Francis Group, an **informa** business

First edition published 2022
by CRC Press
6000 Broken Sound Parkway NW, Suite 300, Boca Raton, FL 33487-2742

and by CRC Press
2 Park Square, Milton Park, Abingdon, Oxon, OX14 4RN

© 2022 Taylor & Francis Group, LLC

CRC Press is an imprint of Taylor & Francis Group, LLC

Reasonable efforts have been made to publish reliable data and information, but the author and publisher cannot assume responsibility for the validity of all materials or the consequences of their use. The authors and publishers have attempted to trace the copyright holders of all material reproduced in this publication and apologize to copyright holders if permission to publish in this form has not been obtained. If any copyright material has not been acknowledged please write and let us know so we may rectify in any future reprint.

Except as permitted under U.S. Copyright Law, no part of this book may be reprinted, reproduced, transmitted, or utilized in any form by any electronic, mechanical, or other means, now known or hereafter invented, including photocopying, microfilming, and recording, or in any information storage or retrieval system, without written permission from the publishers.

For permission to photocopy or use material electronically from this work, access www.copyright. com or contact the Copyright Clearance Center, Inc. (CCC), 222 Rosewood Drive, Danvers, MA 01923, 978-750-8400. For works that are not available on CCC please contact mpkbookspermissions@ tandf.co.uk

Trademark notice: Product or corporate names may be trademarks or registered trademarks and are used only for identification and explanation without intent to infringe.

Library of Congress Cataloging-in-Publication Data
Names: Chu, Liu (Materials scientist), author.
Title: Uncertainty quantification of stochastic defects in materials / Liu Chu.
Description: Boca Raton, FL : CRC Press, 2022. | Series: Emerging materials
  and technologies | Includes bibliographical references and index. |
  Summary: "This book investigates uncertainty quantification methods for
  stochastic defects in material microstructure. Pursuing a comprehensive
  numerical analytical system, the book establishes a fundamental
  framework for this topic, while emphasizing the importance of stochastic
  and uncertainty quantification analysis and the significant influence of
  microstructure defects in the material macro properties. The book is
  intended for researchers, engineers, and advanced students interested in
  material defect quantification methods, material reliability assessment,
  artificial material microstructure optimization, and defect testing"—
  Provided by publisher.
Identifiers: LCCN 2021035516 (print) | LCCN 2021035517 (ebook) |
  ISBN 9781032128733 (hbk) | ISBN 9781032128757 (pbk) |
  ISBN 9781003226628 (ebk)
Subjects: LCSH: Materials—Mathematical models | Quantitative research. |
  Stochastic processes.
Classification: LCC TA404.23.C478 2022 (print) | LCC TA404.23 (ebook) |
  DDC 620.1/1—dc23
LC record available at https://lccn.loc.gov/2021035516
LC ebook record available at https://lccn.loc.gov/2021035517

ISBN: 978-1-032-12873-3 (hbk)
ISBN: 978-1-032-12875-7 (pbk)
ISBN: 978-1-003-22662-8 (ebk)

DOI: 10.1201/9781003226628

Typeset in Times
by codeMantra

This book is a gift for my sons, Hanlin Shi and Chudi Shi.

Contents

SECTION I   Methods and Theories

SECTION II   Examples

Preface

This book investigates the uncertainty quantification methods for the stochastic defects in material microstructures while providing effective supplementary approaches for conventional experimental observation with the consideration of stochastic factors and uncertainty propagation. Studies of stochastic defects quantification methods have attracted engineers and scientists from various disciplines, such as materials, mechanical, electrical, and civil engineering. Pursuing a comprehensive numerical analytical system, the book establishes a fundamental framework for this topic, while emphasizing the importance of stochastic and uncertainty quantification analysis and the significant influence of microstructure defects on the material macro properties. The book is intended for undergraduate and graduate students who are interested in material defect quantification methods and material reliability assessment; researchers investigating the artificial material microstructure optimization; and engineers working on defect influence analysis and nondestructive defect tests.

Thanks for the financial support from the National Natural Science Foundation of China (Grant No. 12102203), the Natural Science Foundation of Jiangsu Province (No. BK20200971), and the Natural Science Foundation of High Education in Jiangsu Province (No. 19KJB130001).

Author

Dr. Liu Chu received her Ph.D. degree in Mechanics from the Institut national des sciences appliquées de Rouen (INSA Rouen), Rouen, France, in 2017, and her B.E. degree in Materials Science and Engineering and M.E. degree in Mechanics from Dalian Maritime University, Dalian, China, in 2010 and 2012, respectively. Dr. Chu focuses on the research of computational material mechanics and structural reliability. Her recent research interests include low-dimensional nanomaterial vacancy defects quantification, artificial material microstructure optimization, and mechanical structure reliability analysis. Since 2018, Dr. Chu has published 21 peer-reviewed scientific and technical papers in international journals. She is a member of IEEE and served as a reviewer of several international journals.

1 Overview

Microdefects in the initial structure of materials are unpredicted and stochastically distributed, which are challenging issues in the experimental and numerical fields.

In the experimental test, microdefects are unstable characteristic existences, which can be disturbed or have dynamic expansion and change according to the slight thermal or pressure fluctuation in the operating environment. Observation and tracking of microdefects in a material are difficult due to the instability. In addition, the scale and size of microdefects are other barriers to the experimental test. Since microdefects are small in magnitude to nm, the requirements for a high resolution of microscopy are very necessary. Moreover, the shape and geometrical configuration of microdefects in graphene are not regular as the often-used geometrical definition. Thus, the location, size, geometrical configuration, and also instability of microdefects are key issues deserving more concern.

In numerical analysis, the appropriate numerical or analytical models for pristine materials can be mainly grouped into categories: tight-binding potentials [1–3], density functional theory (DFT) [4,5], and molecular dynamics (MD) simulation [6,7] and also the finite element method. The introduction of microdefects involves the description of randomly distributed location, scale coordination, irregular geometrical nonlinearity, and also structural instability. Furthermore, the computational cost of the numerical investigation is another issue in the method exploration. Thus, the accuracy for the defect description, computational efficiency and convergence for numerical methods are the crucial issues in the numerical investigation fields of microdefects in materials.

In this book, the quantification methods for the stochastic defects in material microstructures are introduced with verified examples. The stochastic defect quantification methods are the effective supplementary approaches for the conventional experimental observation with the consideration of stochastic factors and uncertainty propagation. Studies on stochastic defect quantification methods have attracted engineers and scientists from various disciplines, such as materials, mechanical, electrical, and civil engineering. Pursuing a comprehensive numerical analytical system, the book establishes a fundamental framework for this topic, while emphasizing the importance of uncertainty quantification analysis and the significant influence of microstructure defects on the material macro properties.

This book consists of two parts: the first part (from Chapters 3 to 6) includes the methods and theories, and the second part (from Chapters 7 to 12) provides the related examples.

Chapter 2 introduces the definition of stochastic defects in materials and presents the uncertainty quantification for the defect location, size, geometrical configuration, and also instability. The impact prediction in mechanical and thermal properties is provided according to the probabilistic statistics.

Chapter 3 introduces the general Monte Carlo (MC) methods, which are widely used for uncertainty quantification with a solid mathematical foundation. In Chapter 3, the mathematical formulation of MC methods is followed by the advanced methods including importance sampling, Latin hypercube sampling, and weight function approaches. Moreover, the random interpolation, iterative MC method, and also Morkov Chain MC method are also presented in Chapter 3.

Chapter 4 mainly introduces the polynomial chaos expansion (PCE). In this chapter, besides the fundamental description of PCE, the stochastic approximation, Hermite polynomial, and Gram–Charlier series are presented. Karhunen–Loève (KL) transform and KL expansion to solve the eigenvalue problem are the foundation of Chapter 5.

In Chapter 5, based on Chapter 4, the stochastic finite element methods are presented. It includes discretization methods, perturbation stochastic finite element method, spectral stochastic finite element, Neumann stochastic finite element method, and finite element reliability analysis.

In Chapter 6, machine learning methods are presented, including the artificial neural network, radial basis network, and back propagation neural network. Moreover, the restricted Boltzmann machine, Hopfield neural network, and the convolutional neural network are also introduced in Chapter 6.

In Chapter 7, numerical examples are provided, including an importance sampling example related to the contents in Chapter 3, and the orthogonal polynomial example related to the PCE in Chapter 4, the Gram–Charlier series as a basic example for Chapter 5, and the Kriging surrogate model for Chapter 6.

In Chapter 8, the Monte Carlo-based finite element method (MC-FEM) as one of the stochastic finite element methods is proposed and simulated to analyze the buckling behavior of vacancy-defected graphene. The critical buckling stress of vacancy-defected graphene sheets deviates within a certain interval. The histogram and regression results of probability density distribution are also given. The strengthening mechanism is detected when the amount of vacancy defects is small. For high-order buckling modes, the symmetrical geometry properties in the vector sum of the displacement are damaged by a large amount of randomly dispersed vacancy defects.

In Chapter 9, different from the molecular dynamic theory and continuum mechanics theory, the MC-FEM is proposed and performed to simulate the vibration behavior of vacancy-defected graphene. Based on the MC simulation, the difficulties in a randomly distributed location of vacancy defects are well overcome. The beam element is chosen to represent the exact atomic lattice of graphene. The results of MC-FEM have a satisfactory agreement with those of the reported references. The natural frequencies in the certain vibration mode are captured to observe the mechanical property of vacancy-defected graphene sheets. The discussion about the parameters corresponding with geometry and material property is accomplished by the probability theory and mathematical statistics.

Due to the inevitable presence of random defects, unpredictable grain boundaries in macroscopic samples, stress concentration at clamping points, and unknown load distribution in the investigation of graphene sheets, uncertainties become a crucial and challenging issue that requires more exploration, which is presented in Chapter 10. The application of the Kriging surrogate model in vibration analysis of graphene sheets is proposed. The Latin hypercube sampling method effectively propagates the uncertainties in geometrical and material properties of the finite element model. The accuracy and convergence of the Kriging surrogate model are confirmed by comparison with the reported references. The uncertainty analyses for both zigzag and armchair graphene sheets are compared and discussed.

In Chapter 11, the uncertain and unavoidable vacancy defects in graphene have an inevitable influence on the extraordinary intrinsic in-plane strength. The equivalent Young's modulus is derived from the strain energy as an important factor to evaluate the stiffness of the entire graphene based on the mechanical molecular theory. The location of vacancy defects in graphene is discussed in the regular deterministic and uncertain patterns. In terms of the boundary condition, shear stress is loaded in the armchair and zigzag edges, respectively. The results show that the center-concentrated vacancy defects deteriorate the elastic stiffness under shear stress. The influences of periodic and regular vacancy defects are sensitive to the boundary condition. By the MC-FEM, vacancy defects are dispersed randomly and propagated. The results of the equivalent Young's modulus are compared with the original values in pristine graphene. The interval and mean values of Young's modulus, total strain, and energy density are also provided and discussed. Compared with the results of graphene with vacancy defects under uniaxial tension, the enhancement effects of vacancy defects are less evident in graphene under shear stress.

In Chapter 12, the possibility of Young's modulus improvement by random vacancy defects is confirmed. The enhancement and optimization of mechanical properties are practical in the design and manufacture process. In this study, MC simulation is used to distribute the random vacancy defects. The uniaxial tension is loaded in the armchair and zigzag edges, respectively. The probability of strengthening effects and maximum Young's modulus with different amounts of randomly distributed vacancy defects are presented and compared.

REFERENCES

1. Mendez, J.P., Ariza, M.P. Harmonic model of graphene based on a tight binding interatomic potential. *Journal of the Mechanics and Physics of Solids*, **2016**, 93, 198–223.
2. Oliveira Neto, P.H., Van Voorhis, T. Dynamics of charge quasiparticles generation in armchair graphene nanoribbons. *Carbon*, **2018**, 132, 352–358.
3. Sinitsa, A.S., et al. Long triple carbon chains formation by heat treatment of graphene nanoribbon: Molecular dynamics study with revised Brenner potential. *Carbon*, **2018**, 140, 543–556.
4. Özkaya, S., Blaisten-Barojas, E. Polypyrrole on graphene: A density functional theory study. *Surface Science*, **2018**, 674, 1–5.
5. Ganji, M.D., Sharifi, N., Ghorbanzadeh Ahangari, M. Adsorption of H_2S molecules on non-carbonic and decorated carbonic graphenes: A van der Waals density functional study. *Computational Materials Science*, **2014**, 92, 127–134.
6. Tsai, J.L., Tu, J.F. Characterizing mechanical properties of graphite using molecular dynamics simulation. *Materials & Design*, **2010**, 31(1), 194–199.
7. Javvaji, B., et al. Mechanical properties of Graphene: Molecular dynamics simulations correlated to continuum based scaling laws. *Computational Materials Science*, **2016**, 125, 319–327.

2 Uncertainty Quantification

2.1 STOCHASTIC DEFECTS

The characteristic description of defects in the internal microstructure of materials includes location, size, geometrical configuration, and amounts. Defects in materials are the inevitable uncertainties in the manufacturing process and operating environment. Defects in materials are unavoidable even in the sophisticated manufacturing process, it is possible to reduce the amount of defects, but it is impossible to 100% eliminate the defects in materials. Moreover, defects as instability existences can be caused by the fluctuation of temperature, force, light, and so on in the operation environment. Therefore, it is crucial to explore stochastic defect quantification methods for the study of specific uncertainties in materials.

The form of stochastic defects in materials includes grain boundaries, vacancy defects, topological defects, and so on. For specific materials, the stochastic defects can be one-dimensional as points, two-dimensional as lines, or three-dimensional as blocks. For typical two-dimensional nanomaterials, the random porosity in graphene [1] is presented in Figure 2.1. The covalent bonds between carbon atoms in graphene ensure the stability in mechanical and chemical properties [2–5]. However, the random porosities are inevitable and significant issues in the research and engineering fields. On the one hand, the atomic [6–8] and bond [9,10] vacancy defects appear

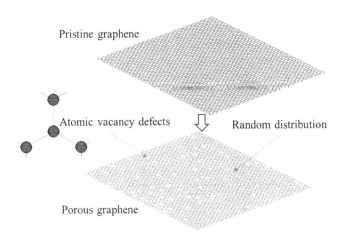

FIGURE 2.1 Schematic of graphene with random porosity (the percentage of atomic vacancy defects is 1.5%).

DOI: 10.1201/9781003226628-2

in the production process of graphene [11–13]. The effects of random porosities in graphene are the important problems that must be confronted. On the other hand, the porosity in graphene not only plays the negative factors in the operating environment, but can also be used and designed to enhance the competence of hydrogen storage and release [14–16], the piezoelectric effects after polarization [17,18], and other positive influences in the applications.

The challenges confronted for the study of the random porosities in graphene are mainly in three aspects. First, the small size in nanometer scale makes the precise measurement in physical experiments difficult and inconvenient, and the experimental equipment is supposed to satisfy more advanced and strict requirements [19–21]. Second, the randomly distributed porosities in graphene contribute to the deviation and variances in the results, no matter which are measured from the experiments or computed in the numerical simulations [22,23]. The confusions in uncertain results of the porous graphene set up obstacles for the comprehensive understanding of the properties of graphene [24]. Third, the general concerns about an independent parameter ignore the correlation and relationships between parameters corresponding to mechanical and physical properties [25]. For example, the resonant frequencies are related to both the mass and stiffness of porous graphene [26]. Therefore, the precise introduction and description of the random porosity in pristine graphene is the premise for further work.

2.2 UNCERTAINTY QUANTIFICATION

According to its specific characteristics, uncertainty should be represented in the research and design process by reasonable approaches. In the different context, model input and model parameter uncertainties have different features. The most widely used research methods include probability theory, evidence theory, possibility theory, interval analysis, and convex modeling [27].

2.2.1 PROBABILITY THEORY

The probability theory is a more prevalent or better-known theory to engineers. Its relative advantages are due to a sound theoretical foundation and long-time development.

In the probability theory, uncertainty is represented as a random variable or a stochastic process. Let X denote the quantity of interest whose probability density function (PDF) is given by $fX(x/p)$, and cumulative distribution function (CDF), where p refers to the distribution parameters of the random variable X (continuous random variable), and x is a realization of X. For a discrete random variable, a sample space is firstly defined which relates to the set of all possible outcomes, each element of the sample space is assigned a probability value between 0 and 1, and the sum of all the elements in the sample space to the "probability" value is called the probability mass function (PMF).

In the context of a probabilistic approach, there are two challenges are encountered. The first is the choice of the distribution type (normal, lognormal, etc. as in Figure 2.2). The choice of the distribution type is known from previous experiences, prior knowledge, or expert opinion, and it is quite subjective. The second difficulty is the lack of sufficient data to estimate the distribution parameters with a high degree of confidence [28].

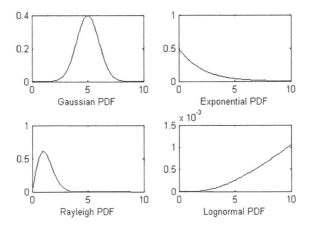

FIGURE 2.2 Examples of the probability density function.

The conventional statistics-based method addresses the uncertainty in the distribution parameters by estimating statistical confidence intervals, which is not feasible or appropriate to be used further in uncertainty propagation, reliability analysis, and so on. In contrast, Bayesian probability interprets the concept of probability as the measurement of a state of belief or knowledge of the quantity of interest, not as a frequency or a physical property of a system. It specifies some prior probability subjectively [29] and then updates in the light of new evidence or observations by means of a statistical inference approach. In this way, it can combine pre-existing knowledge with subsequently available information and update the prior knowledge with uncertainties. With the capability to deal with both aleatory and epistemic uncertainties, the Bayesian theory has been widely applied, especially in reliability engineering.

2.2.2 EVIDENCE THEORY

The evidence theory (Dempster–Shafer theory, D–S theory) measures uncertainty with belief and plausibility determined from known information. For a proposition, lower and upper bounds of probability with consistent evidence are defined instead of assigning a precise probability [30]. The information or evidence to measure belief and plausibility comes from a wide range of sources (e.g., experimental data, theoretical evidence, experts' opinion concerning belief in the value of a parameter or occurrence of an event, and so on).

The evidence theory begins with defining a frame of discernment X, which includes a set of mutually exclusive "elementary" propositions. The elements of the power set $2X$ can be taken to represent propositions concerning the actual state of the system. The evidence theory assigns a belief mass to each element of the power set by a basic belief assignment function m: $2X \rightarrow [0,1]$, which has the following two properties: the mass of the empty set is zero and the mass of all the member elements of the power set adds up to a total of 1.

The mass $m(A)$ expresses the proportion of all relevant and available evidence that supports the claim that the actual state belongs to A. The value of $m(A)$ pertains only to A and makes no additional claims about any subsets of A, each of which has its own mass. From the mass assignments, a probability interval can be defined which contains the precise probability and the lower and upper bound measures are belief (Bel) and plausibility (Pl) as $\mathrm{Bel}(A) \leq P(A) \leq \mathrm{Pl}(A)$.

The belief $\mathrm{Bel}(A)$ is defined as the sum of the mass of all the subsets of A, which represents the amount of all the evidence supporting that the actual state belongs to A, and the plausibility $\mathrm{Pl}(A)$ is the sum of the mass of all the sets that intersect with A, which represents the amount of all the evidence that does not rule out that the actual state belongs to A:

$$\mathrm{Bel}(A) = \sum_{B \,|\, B \in A} m(B) \tag{2.1}$$

$$\mathrm{Pl}(A) = \sum_{B \,|\, B \cap A \neq \varnothing} m(B) \tag{2.2}$$

$$\mathrm{Pl}(A) = \sum_{B \,|\, B \cap A \neq \varnothing} m(B)$$

$$\mathrm{Bel}(A) = \sum_{B \,|\, B \in A} m(B)$$

$$\mathrm{Pl}(A) = 1 - \mathrm{Bel}(\bar{A})$$

$$\mathrm{Bel}(A) + \mathrm{Bel}(\bar{A}) \leq 1$$

$$\mathrm{Pl}(A) + \mathrm{Pl}(\bar{A}) \geq 1$$

The two measures are related to each other as

$$\mathrm{Pl}(A) = 1 - \mathrm{Bel}(\bar{A})$$

$$\mathrm{Bel}(A) + \mathrm{Bel}(\bar{A}) \leq 1 \tag{2.3}$$

$$\mathrm{Pl}(A) + \mathrm{Pl}(\bar{A}) \geq 1$$

where $\bar{A}$ is the complement of A.

The evidence space is characterized by cumulative belief function (CBF) and cumulative plausibility function (CPF).

The evidence theory can deal with the problems both of aleatory and epistemic uncertainties flexibly with its evidence combination rules to update probability measures [32]. It is actually closely related to the probability theory. When the number of available information increases, an uncertainty representation with the evidence theory can approach that with the probability theory [33,34]. However, it also has limitations when handling highly inconsistent data sources, which may render the evidence combination rule unreliable. Anyway, it has been widely utilized and has attracted great research interest in the fields of uncertainty-based information, risk assessment, decision-making, and design optimization [35,36].

2.2.3 Possibility Theory

The possibility theory is introduced as an extension of the theory of fuzzy set and fuzzy logic, which can be used to model uncertainties when there is imprecise information or sparse data. The term fuzzy set is in contrast with the conventional set (fixed boundaries).

In the possibility theory, uncertain parameters are not treated as random variables but as possibilistic variables, and the membership function is extended to possibility distribution.

Like the evidence theory, the possibility theory can deal with both the aleatory and epistemic uncertainties [37]. Compared to the probability theory, the possibility theory can be more conservative in terms of confidence level [38]. The application of a fuzzy set and possibility theory is feasible in engineering design optimization and decision-making. Fractile approach, modality optimization approach, and spread minimization approach can also be developed to solve possibilistic programming problems [39]. The possibility theory can also be applied along with the probability theory, and the integrated or unified algorithms are necessary to research and explore [40–44].

2.2.4 Interval Analysis

Interval analysis is an approach to putting bounds on rounding errors and measurement errors in mathematical computation and yield reliable results by developing numerical methods. In interval analysis, the value of a variable is replaced by a pair of numbers representing the maximum and minimum values that the variable is expected to take. Interval arithmetic rules are applied to perform mathematical operations with the interval numbers, the propagation of the interval bounds through the computational model is implemented, and the bounds on the output variables are achieved [45–48].

2.2.5 Convex Modeling

Convex modeling is a more general approach to represent uncertainties with convex sets [49]. The convex models include the energy-bound model, interval model, ellipsoid model, envelope-bound model, slope-bound model, Fourier-bound model, etc.

Generally, the uncertain components are not independent of each other and the bounds on the components are not reached simultaneously. Therefore, it is more reasonable to apply the convex model with the representation of correlations between uncertain components in a realistic application. In addition, techniques in interval analysis can be used here, when the convex models have intervals [50,51].

Besides the foregoing five theories, there are other alternative approaches to represent uncertainties, especially for epistemic uncertainty [52], such as the cloud theory mediating between the fuzzy set theory and probability distribution, fuzzy random theory, and random fuzzy theory with characteristics of both the fuzzy set theory and probability theory [53,54].

2.3 CORRELATION ANALYSIS

In order to discuss whether a specific distribution is suitable to a dataset, the goodness of fit criteria can be applied, which include the Pearson test [55], the Kolmogorov–Smirnov test [56], the Anderson–Darling test [57], etc.

The estimated correlation matrix is a symmetric matrix of the order N_{var} and can be written as the sum

$$A = I + L + L^T \tag{2.4}$$

where I is the identity matrix and L is the strictly lower triangular matrix with entries in the range $\langle -1,1 \rangle$. There are N_c correlations that describe pairwise correlations:

$$N_c = \binom{N_{\text{var}}}{2} = \frac{N_{\text{var}}(N_{\text{var}} - 1)}{2} \tag{2.5}$$

2.3.1 PEARSON CORRELATION COEFFICIENT

The most well-known correlation measure is the linear Pearson correlation coefficient (PCC) [58]. The PCC takes values between -1 and $+1$, inclusively, and provides an evaluation of the strength of the linear relationship between two variables. The actual PCC between two variables, X_i and X_j, is estimated using the sample correlation coefficient A_{ij} as

$$A_{ij} = \frac{\displaystyle\sum_{s=1}^{N_{\text{sim}}} (x_{i,s} - \overline{X}_i)(x_{j,s} - \overline{X}_j)}{\sqrt{\displaystyle\sum_{s=1}^{N_{\text{sim}}} (x_{i,s} - \overline{X}_i)^2 \sum_{s=1}^{N_{\text{sim}}} (x_{j,s} - \overline{X}_j)^2}} \tag{2.6}$$

$$\overline{X}_i = \frac{1}{N_{\text{sim}}} \sum_{s=1}^{N_{\text{sim}}} x_{i,s}, \quad \overline{X}_j = \frac{1}{N_{\text{sim}}} \sum_{s=1}^{N_{\text{sim}}} x_{j,s} \tag{2.7}$$

When the actual data $x_{i,s}$, $s = 1,2,\ldots,N_{\text{sim}}$ of each vector $i = 1,2,\ldots,N_{\text{var}}$ are standardized into $z_{i,s}$ vectors that yield zero average and unit sample variance estimates, the formula simplifies to

$$A_{ij} = \frac{\sum r_{i,s} r_{j,s}}{\sqrt{\sum r_{i,s}^2 \sum r_{j,s}^2}} \tag{2.8}$$

$$A_{ij} = \frac{1}{N_{\text{sim}}} \sum_{s=1}^{N_{\text{sim}}} z_{i,s} z_{j,s} \tag{2.9}$$

2.3.2 SPEARMAN CORRELATION COEFFICIENT

The formula for Spearman correlation coefficient [59] estimation is identical to the one for Pearson linear correlation with the exception that the values of random variables X_i and X_j are replaced with the ranks $\pi_{i,s}$ and $\pi_{j,s}$, $s = 1, 2, \cdots N_{\text{sim}}$. It is convenient to transform the ranks into $r_{i,s} = \pi_{s,i} - \overline{\pi}_i$ and $r_{j,s} = \pi_{s,j} - \overline{\pi}_j$.

$$\overline{\pi}_i = \overline{\pi}_j = \overline{\pi} = \frac{1}{N_{\text{sim}}} \sum_{s=1}^{N_{\text{sim}}} s = \frac{N_{\text{sim}} + 1}{2} \tag{2.10}$$

The rank correlation is then defined as

$$A_{ij} = \frac{\sum r_{i,s} r_{j,s}}{\sqrt{\sum r_{i,s}^2 \sum r_{j,s}^2}} \tag{2.11}$$

By noting that the sum of the first N_{sim} squared integers is $\dfrac{N_{\text{sim}}(N_{\text{sim}} + 1)(2N_{\text{sim}} + 1)}{6}$, we find that $\sum r_{i,s}^2 = \sum r_{j,s}^2 = \dfrac{N_{\text{sim}}^3 - N_{\text{sim}}}{12}$, and the rank correlation express as

$$A_{ij} = \frac{12 \sum r_{i,s} r_{j,s}}{N_{\text{sim}}(N_{\text{sim}}^2 - 1)} = \frac{12 \sum \pi_{i,s} \pi_{j,s}}{N_{\text{sim}}^3 - N_{\text{sim}}} - 3 \frac{N_{\text{sim}} + 1}{N_{\text{sim}} - 1} \tag{2.12}$$

Note that when Latin Hypercube sampling is applied to continuous parametric distributions, no ties can occur in the generated data. Therefore, the correlation coefficient between any two vectors each consisting of permutations of integer ranks from 1 to N_{sim} is

$$A_{ij} = 1 - \frac{6D}{N_{\text{sim}}(N_{\text{sim}}^2 - 1)} \tag{2.13}$$

where D is the sum of values d_s, the differences between the s th integer elements in the vectors:

$$D = \sum_{s=1}^{N_{\text{sim}}} d_s^2 \tag{2.14}$$

Every mutual permutation of ranks can be achieved by permuting the ranks π_s of the second variable against the identity permutation corresponding to the ranks of the first variable. Therefore,

$$D = \sum_{s=1}^{N_{\text{sim}}} (s - \pi_s)^2 = 2\left[\sum_{s=1}^{N_{\text{sim}}} s^2 - \sum_{s=1}^{N_{\text{sim}}} (s\pi_s) \right] \tag{2.15}$$

This is equal to

$$\frac{N_{\text{sim}}(N_{\text{sim}} + 1)(2N_{\text{sim}} + 1)}{3} - 2\sum_{s=1}^{N_{\text{sim}}} (s\pi_s) \tag{2.16}$$

Spearman correlation depends on the value of the sum $\sum d_s^2$. The lowest correlation is achieved for the reverse ordering of rank numbers and corresponds to the case when the sum D is equal to $\dfrac{N_{\text{sim}}(N_{\text{sim}}^2 - 1)}{3}$. Conversely, the maximum correlation is achieved for identical ranks and the sum equals zero.

2.3.3 KENDALL CORRELATION COEFFICIENT

Kendall's correlation [60] (nonparametric or distribution-free) coefficient estimates the difference between the probability of concordance and discordance between two variables, x_i and x_j. For data without ties, the estimate is calculated based on the rankings π_i and π_j of N_{sim} samples of two vectors x_i and x_j. The ranks are indexed by $1 \leq k, l \leq N_{\text{sim}}$. The formula for sample correlation is a direct estimation of the difference between the probabilities:

$$A_{ij} = \frac{n_c - n_d}{\binom{N_{\text{sim}}}{2}} = \frac{\sum_{k<l}^{N_{\text{sim}}} sng\left[\left(\pi_{i,k} - \pi_{i,l}\right)\left(\pi_{j,k} - \pi_{j,l}\right) \right]}{\binom{N_{\text{sim}}}{2}} \tag{2.17}$$

where sgn(z) = -1 for negative z, $+1$ for positive z, and zero for $z=0$.

The numerator counts the difference between concordant pairs n_c and discordant pairs n_d. The denominator is the maximum number of pairs with the same order, the total number of item pairs with respect to which the ranking can be compared. The number of concordant pairs n_c is the number of item pairs on the order of which both rankings agree. A pair $\left(\pi_{i,k}, \pi_{j,k}\right)$ and $\left(\pi_{i,l}, \pi_{j,l}\right)$ of points in the sample is concordant if either $\pi_{i,k} < \pi_{i,l}$ and $\pi_{j,k} < \pi_{j,l}$ or $\pi_{i,k} > \pi_{i,l}$ and $\pi_{j,k} > \pi_{j,l}$. Analogically, n_d is the number on which both rankings disagreed.

The number of concordant pairs can be calculated by adding scores: a score of one for every pair of objects that are ranked in the same order and a zero score for every pair that is ranked in different orders:

$$n_c = \sum_{k=1}^{N_{\text{sim}}-1} \sum_{l=k+1}^{N_{\text{sim}}} \left(1_{(\pi_{i,k}-\pi_{i,l})(\pi_{j,k}-\pi_{j,l})>0} \right) \tag{2.18}$$

where the indicator function 1_A equals one if A is true and zero otherwise, n_d would count only for opposite orders and the formula would be identical but with an opposite orientation of the inequality sign.

In the cases of tied rank, the denominator is usually adjusted. A_{ij} can be rewritten by exploiting the fact that the number of pairs is the sum of concordant and discordant pairs and therefore the number of discordant pairs is $n_d = \begin{pmatrix} N_{\text{sim}} \\ 2 \end{pmatrix} - n_c$. Then,

$$A_{ij} = \frac{4n_c}{N_{\text{sim}}(N_{\text{sim}}-1)} - 1 = 1 - \frac{4n_d}{N_{\text{sim}}(N_{\text{sim}}-1)} \tag{2.19}$$

A straightforward implementation of the algorithm based on the above equations has $\vartheta(N_{\text{sim}}^2)$ complexity. In practice, it is convenient to rearrange the two rank vectors so that the first one is in increasing order.

Kendall's correlation coefficient is simple to interpret. When compared to the Spearman coefficient, its algebraic structure is much simpler. Note that Spearman's coefficient involves a concordance relationship among three sets of observation, which makes the interpretation somewhat more complex than that for Kendall's coefficient. Regarding the relation between Spearman's correlation (ρ) and Kendall's correlation (τ),

$$\tau - (1 - \tau^2) \le 3\tau - 2\rho \le \tau + (1 - \tau^2) \tag{2.20}$$

For many joint distributions, the correlation coefficients of Spearman and Kendall have different values, as they measure different aspects of the dependence relationship.

2.4 SENSITIVITY ANALYSIS

Sensitivity analysis is the study of how the variation in the model output can be apportioned, qualitatively or quantitatively, to different sources of variations in the model input [61]. Based on the sensitivity analysis, uncertainty factors can be systematically studied to measure their effects on the system output, to filter out the uncertainty factors with negligible contributions and reduce complexity. With this specific aim, sensitivity analysis in this context is also termed uncertainty importance analysis.

There are numerous approaches to address sensitivity analysis under uncertainty, especially with the probability theory. Probabilistic sensitivity analysis methods mainly include differential analysis, response surface method, variance decomposition, Fourier amplitude sensitivity test, sampling-based method [62], etc. Among these approaches, the sampling-based method is widely applied for its flexibility and ease of implementation.

Once the sample is generated, evaluation of f created the following mapping from analysis inputs to analysis results $[x_i, y_i]$, $i = 1, 2, ..., nS$

where $y_i = f(x_i)$, then

$$E(y) = \sum_{i=1}^{nS} y_i w_i \tag{2.21}$$

$$V(y) = \sum_{i=1}^{nS} [E(y) - y_i]^2 w_i \tag{2.22}$$

The mapping $[x_i, y_i]$, $i = 1, 2, ..., nS$ can be explored with various techniques to determine the effects of the individual elements of x on y.

Differential analysis is based on the partial derivative of f with respect to the elements of x. In its simplest form, differential analysis involves approximating the model by the Taylor series

$$y(x) = f(x_0) + \sum_{j=1}^{nX} [\partial f(x_0)/\partial x_j][x_j - x_{j0}] \tag{2.23}$$

where $x_0 = [x_{10}, x_{20}, ..., x_{nX,0}]$ is a vector of base-case values for x_j.

Once the approximation in the model of the Taylor series is determined, variance propagation formulas can be used to determine the uncertainty in y that results from the distribution. In particular,

$$E(y) = y(x_0) + \sum_{j=1}^{nX} [\partial f(x_0)/\partial x_j] E[x_j - x_{j0}] \tag{2.24}$$

$$V(y) = \sum_{j=1}^{nX} [\partial f(x_0)/\partial x_j]^2 V(x_j) + 2 \sum_{j=1}^{nX} \sum_{k=j+1}^{nX} [\partial f(x_0)/\partial x_j] \times [\partial f(x_0)/\partial x_k] Cov(x_j, x_k)$$

$$\tag{2.25}$$

Thus, the Taylor series leads to approximations of the expected value and variance for y that result from the distributions. Sensitivity analysis is based on the use of partial derivatives associated with a Taylor series to determine the effects of the individual elements. If the elements are independent, then the fractional contribution of x_j to the variance of y can be approximated by

$$V(y/x_j) = [\partial f(x_0)/\partial x_j]^2 V(x_j)/V(y) \tag{2.26}$$

REFERENCES

1. Novoselov, K.S., et al. Electric field effect in atomically thin carbon films. *Science*, **2004**, 306(5696), 666–669.
2. Bonilla, L.L., Carpio, A. Theory of defect dynamics in graphene: Defect groupings and their stability. *Continuum Mechanics and Thermodynamics*, 23(4), **2011**, 337–346.
3. Kim, H.S., Oweida, T.J., Yingling, Y.G. Interfacial stability of graphene-based surfaces in water and organic solvents. *Journal of Materials Science*, **2018**, 53(8), 5766–5776.
4. Ariza, M.P., Ortiz, M., Serrano, R. Long-term dynamic stability of discrete dislocations in graphene at finite temperature. *International Journal of Fracture*, **2010**, 166(1–2), 215–223.
5. Rani, P., Jindal, V.K. Stability and electronic properties of isomers of B/N co-doped graphene. *Applied Nanoscience*, **2014**, 4(8), 989–996.
6. Nayebi, P., Zaminpayma, E., Emami-Razavi, M. Study of electronic properties of graphene device with vacancy cluster defects: A first principles approach. *Thin Solid Films*, **2018**, 660, 521–528.
7. Li, T., Yarmoff, J.A. Defect-induced oxygen adsorption on graphene films. *Surface Science*, **2018**, 675, 70–77.
8. Araujo, E.N.D., et al. Quantum corrections to conductivity in graphene with vacancies. *Physica E: Low-Dimensional Systems and Nanostructures*, **2018**, 100, 40–44.
9. Son, J., et al. Structural evolution of graphene in air at the electrical breakdown limit. *Carbon*, **2016**, 99, 466–471.
10. Okada, T., et al. Bonding state and defects of nitrogen-doped graphene in oxygen reduction reaction. *Chemical Physics Letters*, **2016**, 665, 117–120.
11. Geim, A.K. Graphene: Status and prospects. *Science*, **2009**, 324(5934), 1530–1534.
12. Grantab, R., Shenoy, V.B., Ruoff, R.S. Anomalous strength characteristics of tilt grain boundaries in graphene. *Science*, **2010**, 330(6006), 946–948.
13. Terdalkar, S.S., et al. Nanoscale fracture in graphene. *Chemical Physics Letters*, **2010**, 494(4–6), 218–222.
14. Tozzini, V., Pellegrini, V. Reversible hydrogen storage by controlled buckling of graphene layers. *The Journal of Physical Chemistry C*, **2011**, 115(51), 25523–25528.
15. Roszak, R., et al. Hydrogen storage by adsorption in porous materials: Is it possible? *Colloids and Surfaces A: Physicochemical and Engineering Aspects*, **2016**, 496, 69–76.
16. Yadav, S., Zhu, Z., Singh, C.V. Defect engineering of graphene for effective hydrogen storage. *International Journal of Hydrogen Energy*, **2014**, 39(10), 4981–4995.
17. Hinchet, R., et al. Piezoelectric properties in two-dimensional materials: Simulations and experiments. *Materials Today*, **2018**, 21(6), 611–630.
18. Kundalwal, S.I., Meguid, S.A., Weng, G.J. Strain gradient polarization in graphene. *Carbon*, **2017**, 117, 462–472.
19. Lee, C., et al. Measurement of the elastic properties and intrinsic strength of monolayer graphene. *Science*, **2008**, 321(5887), 385–388.
20. Eckmann, A., et al. Probing the nature of defects in graphene by Raman spectroscopy. *Nano Letters*, **2012**, 12(8), 3925–3930.
21. Ferrari, A.C., et al. Raman spectrum of graphene and graphene layers. *Physical Review Letters*, **2006**, 97(18), 187401.
22. Qin, H., et al. Mechanical properties of wrinkled graphene generated by topological defects. *Carbon*, 2016, 108, 204–214.
23. Chu, L., Shi, J., Ben, S. Buckling analysis of vacancy-defected graphene sheets by the Stochastic finite element method. *Materials*, **2018**, 11(9), 1545.
24. Deng, S., Berry, V. Wrinkled, rippled and crumpled graphene: An overview of formation mechanism, electronic properties, and applications. *Materials Today*, **2016**, 19(4), 197–212.

25. Zandiatashbar, A., et al. Effect of defects on the intrinsic strength and stiffness of graphene. *Nature Communications*, **2014**, 5, 3186.

26. Chu, L., et al. Monte Carlo-based finite element method for the study of randomly distributed vacancy defects in graphene sheets. *Journal of Nanomaterials*, 2018, 2018: 3037063.

27. Haldar, A., Mahadevan, S., *Probability, Reliability, and Statistical Methods in Engineering Design*, John Wiley & Sons Inc. 2000. ISBN: 978-0-471-33119-3.

28. Pawitan, Y., *In All Likelihood: Statistical Modeling and Inference Using Likelihood*, USA: Oxford University Press. 2001. ISBN: 978-0199671229.

29. Youn, B.D., Kloess, A., Bayesian reliability analysis with evolving, insufficient, and subjective data sets. *Journal of Mechanical Design*, **2009**; 131(11), 111008.

30. Yager, R., Kacprzy, K.J., Fedrizzi, M. *Advances in the Dempster-Shafer Theory of Evidence*, New York: John Wiley & Sons Inc. 1994.

31. Sentz, K., Ferson, S. Combination of evidence in Dempster-Shafer theory, SAND 2002-0835, Sandia National Laboratories; 2002.

32. Soundappan, P., Grandhi, R., Canfield, R.. Comparison of evidence theory and Bayesian theory for uncertainty modeling, *Reliability Engineering & System Safety*, **2004**, 85, 295–311.

33. Alyanak, E., Grandhi, R., Bae, H.R. Gradient projection for reliability-based design optimization using evidence theory, *Engineering Optimization*, **2008**, 40(10), 923–935.

34. Bai, Y.C., Jiang, C. Evidence-theory-based structural static and dynamic response analysis under epistemic uncertainties. *Finite Elements in Analysis and Design*, **2013**, 68, 52–62.

35. Agarwal, H., Renaud, J.E., Preston, E.L. Uncertainty quantification using evidence theory in multidisciplinary design optimization. *Reliability Engineering & System Safety*, **2004**, 85, 281–294.

36. Mourelatos, Z.P., Zhou, J. A design optimization method using evidence theory. *Journal of Mechanical Design*, **2006**, 128, 270–279.

37. Jensen, H.A. Structural optimal design of systems with possibility theory. *Advances in Engineering Software*, **2001**, 32, 937–948.

38. Lopez, I., Sarigul-Klijn, N. A review of uncertainty in flight vehicle structural damage monitoring, diagnosis and control: Challenges and opportunities. *Progress in Aerospace Sciences*, **2010**, 46, 247–273.

39. Inuiguchi, M., Ichihashi, H. Modality constrained programming problems: A unified approach to fuzzy mathematical programming problems in the setting of possibility theory. *Information Sciences*, **1993**, 67, 93–126.

40. Mauris, G. Possibility distributions: A unified representation of usual direct-probability-based parameter estimation methods. *International Journal of Approximate Reasoning*, **2011**, 52, 1232–1242.

41. Sakallı, Ü.S., Baykoç, Ö.F. Can the uncertainty in brass casting blending problem be managed A probability/possibility transformation approach. *Computers & Industrial Engineering*, 2011, 61, 729–738.

42. Miranda, E., Destercke, S. On the connection between probability boxes and possibility measures. *Information Sciences*, **2013**, 224, 88–108.

43. Dubois, D., Prade, H. Bayesian conditioning in possibility theory. *Fuzzy Sets and Systems*, **1997**, 92, 223–240.

44. Baudrit, C., Couso, I., Dubois, D. Joint propagation of probability and possibility in risk analysis: Towards a formal framework. *International Journal of Approximate Reasoning*, **2007**, 45, 82–105.

45. Impollonia, N., Muscolino, G. Interval analysis of structures with uncertain-but-bounded axial stiffness. *Computer Methods in Applied Mechanics and Engineering*, **2011**, 200, 1945–1962.

46. Jiang, C., Han, X., Guan, F.J. An uncertain structural optimization method based on nonlinear interval number programming and interval analysis method. *Engineering Structures*, **2007**, **29**, 3168–3177.
47. Majumder, L., Rao, S.S. Interval-based optimization of aircraft wings under landing loads. *Computers & Structures*, **2009**, 87, 225–235.
48. Verhaeghe, W., Desmet, W., Vandepitte, D., Moens, D. Interval fields to represent uncertainty on the output side of a static FE analysis. *Computer Methods in Applied Mechanics and Engineering*, **2013**, 260, 50–62.
49. Ben-Haim, Y., Elishakoff, I. *Convex Models of Uncertainty in Applied Mechanics*, Elsevier. 1990- Psychology, 221 pages.
50. Ben-Haim, Y. Convex models of uncertainty: Applications and implications. *Erkenntnis: An International Journal of Analytic Philosophy*, **1994**, 41, 139–156.
51. Fuchs, M., Neumaier, A. Uncertainty modeling with clouds in autonomous robust design optimization, REC 2008, USA.
52. Troffaes, M., Walter, G., Kelly, D. A robust Bayesian approach to modeling epistemic uncertainty in common-cause failure models. *Reliability Engineering & System Safety*, **May 2014**, 125, 13–21.
53. Helton, J.C., Johnson, J.D. Quantification of margins and uncertainties: Alternative representations of epistemic uncertainty. *Reliability Engineering & System Safety*, **September 2011**, 96(9), 1034–1052.
54. Helton, J.C., Johnson, J.D., Oberkampf, W.L. An exploration of alternative approaches to the representation of uncertainty in model predictions. *Reliability Engineering & System Safety*, **July–September 2004**, 85(1–3), 39–71.
55. Pearson, K. Notes on regression and inheritance in the case of two parents. *Proceedings of the Royal Society of London*, **1895**, 58, 240–242.
56. Shorack, G.R., Wellner, J.A. *Empirical Processes with Applications to Statistics*, Philadelphia: Society for Industrial & Applied Mathematics. 4 September 2009, 998 p. (ISBN 978-0-89871-684-9, LCCN 2009025143).
57. Stephens, M.A. EDF statistics for goodness of fit and some comparisons. *Journal of the American Statistical Association*, **1974**, 69, 730–737.
58. Galton, F. Regression towards mediocrity in hereditary stature. *Journal of the Anthropological Institute of Great Britain and Ireland*, **1886**, 15, 246–263.
59. Lehman, A. *JMP for Basic Univariate and Multivariate Statistics: A Step-by-step Guide*. Cary, NC: SAS Institute. 2005. ISBN: 978-1590475768
60. Kruskal, W.H. Ordinal measures of association. *Journal of the American Statistical Association*, **1958**, 53(284), 814–861.
61. Saltelli, A., Andres, T.H., Homma, T. Sensitivity analysis of model output: An investigation of new techniques. *Computational Statistics & Data Analysis*, **February 1993**, 15(2), 211–238.
62. Helton, J.C., Bean, J.E., Butcher, B.M., Garner, J.W., Schreiber, J.D., Swift, P.N., Vaughn, P. Uncertainty and sensitivity analysis for gas and brine migration at the Waste Isolation Pilot Plant: Permeable shaft with panel seals. *Journal of Hazardous Materials*, **February 1996**, 45(2–3), 107–139.

Section I

Methods and Theories

3 Monte Carlo Methods

3.1 INTRODUCTION

Monte Carlo (MC) methods are a broad class of computational algorithms that rely on repeated random sampling to obtain numerical results. They are often used in physical and mathematical problems for optimization, numerical integration, and stochastic sampling from a probability distribution. MC methods vary, but tend to follow a particular pattern:

1. Defining a domain of possible inputs;
2. Generating inputs randomly from a probability distribution over the domain;
3. Performing a deterministic computation on the inputs;
4. Aggregating the results.

The computational accuracy and convergence of MC methods are highly dependent on the amount of random numbers. In principle, MC methods are feasible to solve a problem with probabilistic interpretation based on the law of large numbers. When the probability distribution of variables is parametrized, a Markov chain Monte Carlo (MCMC) sampler is performed. Using the ergodic theorem, the stationary distribution is approximated by the empirical measures of the random states of the MCMC sampler. Furthermore, when the problem is more complicated for a sequence of probability distributions with the nonlinear relationships, these flows of probability distributions can always be interpreted as the distributions of the random states of a Markov process whose transition probabilities depend on the distributions of the current random states. The nonlinear Markov processes are simulated by multiple copies of the sampling process, replacing in the evolution equation the unknown distributions of the random states by the sampled empirical measures. Each of the samples interacts with the empirical measures of the process. When the size of the system tends to infinity, these random empirical measures converge to the deterministic distribution of the random states of the nonlinear Markov chain, so that the statistical interaction between samples vanishes.

3.1.1 MATHEMATICAL FORMULATION OF MC INTEGRATION

Let ξ be a random variable for which the mathematical expectation $E(\xi) = I$ exists. It is formally defined as

$$E(\xi) = \begin{cases} \displaystyle\int_{-\infty}^{\infty} \xi p(\xi) d\xi & \text{where } \int_{-\infty}^{\infty} p(x) dx = 1 \quad \text{when } \xi \text{ is a continuours r.v.} \\ \Sigma_\xi \xi p(\xi) & \text{where } \Sigma_x p(x) = 1 \quad \text{when } \xi \text{ is a discrete r.v.} \end{cases} \tag{3.1}$$

DOI: 10.1201/9781003226628-4

The nonnegative function $p(x)$ (continuous or discrete) is called the probability density function. To approximate the variable I, a computation of the arithmetic mean must usually be carried out:

$$\bar{\xi}_N = \frac{1}{N}\sum_{i=1}^{N}\xi_i \tag{3.2}$$

For a sequence of uniformly distributed independent random variables, the arithmetic mean of these variables converges to the mathematical expectation:

$$\xi_N \xrightarrow{p} I \quad \text{as } N \to \infty$$

The sequence of the random variables $\eta_1,\eta_2,\dots,\eta_N,\dots$ converges to the constant c if, for every $h > 0$, it follows that

$$\lim_{N\to\infty} P\{|\eta_N - c| \ge h\} = 0$$

Thus, when N is sufficiently large, $\bar{\xi}_N \approx I$. Suppose that the random variable ξ has a finite variance, the error of the algorithm can be estimated as

$$D(\xi) = E[\xi - E(\xi)]^2 = E(\xi^2) - [E(\xi)]^2 \tag{3.3}$$

3.1.2 PLAIN (CRUDE) MC ALGORITHM

Crude MC is the simplest possible approach for solving multidimensional integrals. This approach simply applied the definition of the mathematical expectation. Let Ω be an arbitrary domain and $x \in \Omega \subset R^d$ be a d-dimensional vector.

Consider the problem of the approximate computation of the integral

$$I = \int_{\Omega} f(x)p(x)\,dx \tag{3.4}$$

where the nonnegative function $p(x)$ is the density function $\int_{\Omega} p(x)dx = 1$.

Let ξ be a random point with a probability density function $p(x)$. Introduce the random variable

$$\theta = f(\xi) \tag{3.5}$$

with the mathematical expectation equals to the value of integral I,

$$E(\theta) = \int_{\Omega} f(x)p(x)\,dx \tag{3.6}$$

Let the random points $\xi_1,\xi_2,\dots,\xi_N$ be independent realizations of the random point ξ with probability density function $p(x)$, then an approximate value of I is

$$\bar{\theta}_N = \frac{1}{N} \sum_{i=1}^{N} \theta_i \qquad (3.7)$$

If $\bar{\xi}_N = \frac{1}{N} \sum_{i=1}^{N} \xi_i$ were absolutely convergent, then $\bar{\theta}_N$ would be convergent to I.

3.1.3 GEOMETRIC MC ALGORITHM

Let the nonnegative function f be bounded,

$$0 \leq f(x) \leq c \quad \text{for} \quad x \in \Omega$$

where c is a generic constant. Consider the cylindrical domain $\tilde{\Omega} = \Omega \times [0, c]$ and the random point $\tilde{\xi} = (\xi_1, \xi_2, \xi_3) \subset \tilde{\Omega}$ with the following probability density function

$$\tilde{p}(x) = \tfrac{1}{c} p(x_1, x_2) \qquad (3.8)$$

Let $\tilde{\xi}_1, \dots, \tilde{\xi}_N$ be an independent realization of the random point $\tilde{\xi}$. Introduce the random variable

$$\tilde{\theta} = \begin{cases} c, \text{if } \xi_3 < f(\xi_1, \xi_2) \\ 0, \text{if } \xi_3 < f(\xi_1, \xi_2) \end{cases} \qquad (3.9)$$

The random variable introduced is a measure of the points below the graph of the function f.

$$E(\tilde{\theta}) = c \Pr\{\xi_3 < f(\xi)\}$$

$$= \int_\Omega dx_1 dx_2 \int_0^{f(x_1, x_2)} \tilde{p}(x_1, x_2, x_3) dx_3 = I \qquad (3.10)$$

Compare the accuracy of the geometric and the plain MC algorithm.

Let $f \in L_2(\Omega, p)$ guarantee that the variance $D(\theta) = \int_\Omega f^2(x) p(x) dx - I^2$ in a plain MC algorithm is finite.

For the geometric MC algorithm, the following equation holds

$$E(\tilde{\theta}^2) = c^2 P\{x_3 < f(\xi)\} = cI \qquad (3.11)$$

Hence the variance is $D(\tilde{\theta}) = cI - I^2$. Because

$$\int_\Omega f^2(x) p(x) dx \leq c \int_\Omega f(x) p(x) dx = cI \qquad (3.12)$$

Therefore, $D(\theta) \le D(\tilde{\theta})$. The last inequality shows that the plain MC algorithm is more accurate than the geometric one (except for the case when the function f is a constant). Nevertheless, the geometric algorithm is often preferred, since its computational complexity may be less than that of the plain algorithm.

3.2 ADVANCED MC METHODS

The probable relative errors in MC algorithms will appear when no information about the smoothness of the function is used:

$$r_N = c \sqrt{\frac{D\xi}{N}} \tag{3.13}$$

It is important for such computational schemes and random variables that a value of ξ is chosen so that the variance is as small as possible. MC algorithms with reduced variance compared to plain MC algorithms are usually called advanced MC algorithms. Consider the integral,

$$I = \int_\Omega f(x)p(x)\,dx \tag{3.14}$$

where $f \in L_2(\Omega, p)$, $x \in \Omega \subset R^d$. Let the function $h(x) \in L_2(\Omega, p)$ be close to $f(x)$ with respect to its L_2 norm; $\|f - h\|_{L_2} \le \varepsilon$. The value of the integral is supposed to be

$$I = \int_\Omega h(x)p(x)\,dx = I' \tag{3.15}$$

The random variable $\theta' = f(\xi) - h(\xi) + I'$ generates the following estimate for the integral

$$\theta'_N = I' + \frac{1}{N}\sum_{i=1}^{N}[f(\xi_i) - h(\xi_i)] \tag{3.16}$$

A possible estimate of the variance of θ' is

$$D(\theta') = \int_\Omega \left[f(x) - h(x)\right]^2 p(x)\,dx - (I - I')^2 \le \varepsilon^2 \tag{3.17}$$

This means that the variance and the probable error will be quite small if the integral I' can be calculated analytically. The function $h(x)$ is often chosen to be a piece-wise linear function to compute the value of I' easily.

3.2.1 IMPORTANCE SAMPLING ALGORITHM

Importance sampling is a variance reduction technique that can be used in the MC method. The basic idea behind importance sampling is that certain values of the input random variables in a simulation have more impact on the parameter being estimated than others. If these "important" values are emphasized by sampling more frequently, then the estimator variance can be reduced. Hence, the basic concepts of the importance sampling is to choose a distribution which "encourages" the important values.

Consider the problem of computing the integral,

$$I_0 = \int_\Omega f(x)\,dx, \quad x \in \Omega \subset R^d$$

Ω_0 is the set of points x for which $f(x) = 0$ and $\Omega_+ = \Omega - \Omega_0$.

Define the probability density function $p(x)$ to be tolerant to $f(x)$, if $p(x) > 0$ for $x \in \Omega_+$, and $p(x) \ge 0$ for $x \in \Omega_0$.

For an arbitrary tolerant probability density function $p(x)$ for $f(x)$ in Ω, the random variable θ_0 can be defined in the following way:

$$\theta_0(x) = \begin{cases} \dfrac{f(x)}{p(x)}, & x \in \Omega_+ \\ 0, & x \in \Omega_0 \end{cases} \tag{3.18}$$

It is interesting to consider the problem of finding a tolerant density, $p(x)$, which minimizes the variance of θ_0.

In importance sampling, a distribution g which is called importance distribution or instrumental distribution is introduced.

$$E(\theta) = \mu = \int \frac{p(x)}{g(x)} f(x) g(x)\,dx \tag{3.19}$$

In sampling space, $w(x_i)$ represents important weights:

$$w(x_i) = \frac{p(x_i)}{g(x_i)} \quad i = 1,\dots,n \tag{3.20}$$

$$E(\theta) = \int f(x) p(x)\,dx$$

$$= \int \cdots \int \left(\frac{1}{n} \sum_{i=1}^{n} w(x_i) f(x_i) \right) \prod_{i=1}^{n} g(x_i)\,dx_i \tag{3.21}$$

Importance sampling methods are frequently used to estimate posterior densities or expectations in state and/or parameter estimation problems in probabilistic models that are too hard to treat analytically, for example in Bayesian networks. It is used to estimate properties of a particular distribution, while only having samples generated from a different distribution rather than the distribution of interest.

3.2.2 WEIGHT FUNCTION APPROACH

If the integrand contains a weight function, in MC quadrature, the weight functions are considered for the computation of

$$S(g;m) = \int g(\theta)m(\theta)d(\theta) \tag{3.22}$$

The un-normalized posterior density m is expressed as the product of two functions w and f, where w is called the weight function $m(\theta) = w(\theta)f(\theta)$. The weight function is nonnegative and $\int w(\theta)d\theta = 1$, and it is chosen to have similar properties to m. Most numerical integration algorithms then replace the function $m(\theta)$ by a discrete approximation in the form of:

$$\hat{m}(\theta) = \begin{cases} w_i f(\theta), & \theta = \theta_i, i = 1,2,...n, \\ 0 & \text{elsewhere} \end{cases} \tag{3.23}$$

Then, the integral can be estimated using the following formula:

$$\hat{S}(g;m) = \sum_{i=1}^{N} w_i f(\theta_i)g(\theta_i) \tag{3.24}$$

Integration algorithms use the weight function w as the kernel of the approximation of the integrand

$$S(g;m) = \int g(\theta)m(\theta)d\theta$$

$$= \int g(\theta)w(\theta)f(\theta)d\theta = E[w(g(\theta)f(\theta))] \tag{3.25}$$

3.2.3 LATIN HYPERCUBE SAMPLING APPROACH

The probability error usually has the form of $R_N = cN^{-0.5}$. The speed of convergence can be increased if an algorithm with a probability error $R_N = cN^{-0.5-\psi(d)}$ can be constructed, where c is a constant, $\psi(d) > 0$, and d is the dimension of the space. Usually, the exploitation of smoothness is combined with subdividing the domain of integration into several nonoverlapping sub-domains. This is the reason to call the techniques leading to super-convergent MC algorithms stratified sampling or Latin hypercube sampling (LHS).

LHS also known as the "stratified sampling technique" represents a multivariate sampling method that guarantees a non-overlapping design. In LHS, the distribution for each random variable can be subdivided into n equal probability intervals or bins. Each bin has one analysis point. There are n analysis points, randomly mixed, so each of the n bins has 1/n of the distribution probability. The basic steps are

1. Dividing the distribution for each variable into n nonoverlapping intervals on the basis of equal probability;
2. Selecting one value at random from each interval with respect to its probability density;
3. Repeating steps (1) and (2) until you have selected values for all random variables, such as $x_1, x_2, \ldots, x_k$;
4. Associating the n values obtained for each x_i with the n values obtained for the other $x_{j \neq i}$ at random.

The regularity of probability intervals on the probability distribution function ensures that each of the input variables has all portions of its range represented, resulting in a relatively small variance in the response. Meanwhile, the analysis is much less computationally expensive to generate. The LHS method also provides flexible sample size while ensuring stratified sampling, and each of the input variables is sampled at n levels [2].

3.3   RANDOM INTERPOLATION QUADRATURE

A quadrature is called interpolation for a given class of functions if it is exact for any linear combination of functions. In practical computations, the probability relative errors exist but with small values. However, the problem of the restricted application is more evident. Since the system of orthonormal functions are combined in the linear way. The quadratures are effective to compute the multidimensional integrals with similar integrands.

Assume that the quadrature formula for computing the integral,

$$I = \int_{\Omega} f(x)p(x)dx, \Omega \subset R^d, \, p(x) \geq 0, \int_{\Omega} p(x)dx = 1 \tag{3.26}$$

is denoted by the expression

$$I \approx \sum_{j=1}^{N} c_j f(x_j) \tag{3.27}$$

where $x_1, \ldots, x_N \in \Omega$ are nodes. Then the random quadrature formula can be written in the following form:

$$I \approx \sum_{j=1}^{N} \kappa_j f(\xi_j) \tag{3.28}$$

where $\xi_1, \ldots, \xi_N \in \Omega$ are random nodes and $\kappa_1, \ldots, \kappa_N$ are random weights.

All functions considered are supposed to be partially continuous and belong to the space $L_2(\Omega)$. Let $\varphi_0, \varphi_1, \ldots, \varphi_m$ be a system of orthonormal functions, such that

$$\int_{\Omega} \varphi_k(x)\varphi_j(x)dx = \delta_{kj} \tag{3.29}$$

where δ_{kj} is the Kronecker function [3].

For $p(x) = \varphi_0(x)$, an approximate solution for the integral is

$$I = \int_{\Omega} f(x)\varphi_0(x)\,dx \tag{3.30}$$

Let us fix arbitrary nodes and choose the weight $c_0, c_1, \ldots, c_m$ such that it is exact for the system of orthonormal functions $\varphi_0, \varphi_1, \ldots, \varphi_m$. In this case, it is convenient to represent the quadrature formula as a ratio of two determinants.

$$I \approx \frac{W_f(x_0, x_1, \ldots, x_m)}{W_{\varphi_0}(x_0, x_1, \ldots, x_m)} \tag{3.31}$$

where

$$W_g(x_0, x_1, \ldots, x_m) = \begin{vmatrix} g(x_0) & \varphi_1(x_0) & \varphi_m(x_0) \\ g(x_1) & \varphi_1(x_1) & \varphi_m(x_1) \\ g(x_m) & \varphi_1(x_m) & \varphi_m(x_0) \end{vmatrix} \tag{3.32}$$

It is easy to check that if $W_{\varphi_0} \neq 0$, then the formula is exact for every linear combination of the following form: $f = a_0\varphi_0 + \cdots + a_m\varphi_m$.

3.4   ITERATIVE MC METHODS FOR LINEAR EQUATIONS

In general, MC numerical algorithms may be divided into two classes—direct algorithms and iterative algorithms. Direct algorithms provide an estimate of the solution of the equation in a finite number of steps and contain only a stochastic error. However, iterative MC algorithms deal with an approximated solution in each step of the algorithm.

Iterative algorithms are preferred for solving integral equations and large sparse systems of algebraic equations. In the diagonally dominant systems, the iterative algorithms present the rapid convergence, while in the problems involving dense matrices, the merits of the iterative algorithms are not evident.

Define an iteration of degree j as

$$u^{k+1} = F_k(A, b, u^k, u^{k-1}, \ldots, u^{k-j+1}) \tag{3.33}$$

where u^k is obtained from the k th iteration. Usually, the degree of j is kept small because of storage requirements. When $F_k = F$ for all k, the computation method is stationary iteration

The iterative MC process is performed linearly if F_k is a linear function of $u^k, u^{k-1}, \ldots, u^{k-j+1}$. In the computation accuracy discussion, both systematic and stochastic errors are necessary to be taken into consideration.

In addition, the iterative stationary linear MC algorithms are also called as power MC algorithms, since the approximation of the functional powers of linear operators are performed. Furthermore, the Markov chain MC can be also considered as weights of Markov chains for the same reason.

3.4.1 ITERATIVE MC ALGORITHMS

Consider a general description of the iterative MC algorithms. Let X be a space of real-valued functions. Let $f = f(x) \in X$ and $u_k = u(x_k) \in X$ be defined in Rd and $L = L(u)$ be a linear operator defined on X.

Consider the sequence $u_1, u_2, \ldots$, defined by the recursion formula

$$u_k = L(u_{k-1}) + f, \quad k = 1, 2, \ldots \tag{3.34}$$

The formal solution of this equation is the truncated Neumann series

$$u_k = f + L(f) + \cdots + L^{k-1}(f) + L^k(u_0), \quad k > 0 \tag{3.35}$$

where L^k means the k th iterate of L.

Let $u(x) \in X$, $x \in \Omega \subset R^d$, and $l(x, x')$ be a function defined for $x \in \Omega, x' \in \Omega$. The integral transformation

$$L(u(x)) = \int_\Omega l(x, x') u(x') dx'$$ maps the function to the function $L(u(x))$ and is called an iteration of $u(x)$ by the integral transformation kernel $l(x, x')$. The second integral iteration of $u(x)$ is denoted by

$$L(L(u(x))) = L^2(u(x)) \tag{3.36}$$

Obviously,

$$L^2(u(x)) = \int_\Omega \int_\Omega l(x, x') l(x', x'') dx' dx'' \tag{3.37}$$

In this way, $L^3(u(x)), \ldots, L^i(u(x)), \ldots$ can be defined.

When the infinite series converges, the sum is an element u from the space X, which satisfies the equation:

$$u = L(u) + f \tag{3.38}$$

The truncation error is

$$u_k - u = L^k(u_0 - u) \tag{3.39}$$

Let $J(u_k)$ be a linear functional that is to be calculated. Consider the spaces

$$T_{i+1} = \underbrace{R^d \times R^d \times \cdots \times R^d}_{i \text{ times}}, \quad i = 1, 2, \ldots, k$$

where "×" denotes the Cartesian product of spaces.

Random variables θ_i, $i = 0,1,\dots,k$ are defined on the respective product spaces T_{i+1} and have conditional mathematical expectations:

$$E(\theta_0) = J(u_0),\ E\left(\theta_1/\theta_0\right) = J(u_1),\dots,E\left(\theta_k/\theta_0\right) = J(u_k) \tag{3.40}$$

where $J(u)$ is a linear functional of u.

The computational problem then becomes one of calculating repeated realizations of θ_k and combining them into an appropriate statistical estimator of $J(u_k)$.

As an approximate value of the linear functional, $J(u_k)$ is set up

$$J(u_k) \approx \frac{1}{N}\sum_{s=1}^{N}\{\theta_k\}_s \tag{3.41}$$

where $\{\theta_k\}_s$ is the s th realization of the random variables.

The probable error of the above equation is

$$r_N = c\sigma(\theta_k)N^{-0.5} \tag{3.42}$$

where $c \approx 0.6745$ and $\sigma(\theta_k)$ is the standard deviation of the random variable θ_k

There are two approaches which correspond with two special cases of the operator L:

- L is a matrix, u and f are vectors;
- L is an ordinary integral transform.

$$L(u) = \int_\Omega l(x,y)u(y)\,dy \tag{3.43}$$

First, consider the case of an ordinary integral transform,

$$u(x) = \int_\Omega l(x,y)u(y)\,dy + f(x) \ \text{ or } \ u = L(u) + f \tag{3.44}$$

MC algorithms frequently involve the evaluation of linear functions of the solution of the following type,

$$J(u) = \int_\Omega h(x)u(x)\,dx = (u,h) \tag{3.45}$$

This equation defines an inner product of a given function $h(x) \in X$ with the solution of the integral equation.

Sometimes, the adjoined equation $v = L^*v + h$ will be used.

$v, h \in X^*$, $L^* \in [X \to X^*]$, where X^* is the dual functional space to X and L^* is an adjoined operator.

For some important applications $X = L_1$,

$$\|f\|_{L_1} = \int_\Omega |f(x)|\,dx \tag{3.46}$$

$$\|L\|_{L_1} \le \sup_x \int_\Omega |l(x,x')|\,dx' \tag{3.47}$$

In this case, $h(x) \in L_\infty$, hence $L_1^* \equiv L_\infty$,

$$\|h\|_{L_\infty} = \sup|h(x)|, \, x \in \Omega \tag{3.48}$$

For many applications, $X = X^* = L_2$. Note also that if $h(x), u(x) \in L_2$, then the inner product is finite. In fact,

$$\left|\int_\Omega h(x)u(x)\,dx\right| \le \int_\Omega |h(x)u(x)|\,dx \le \left\{\int_\Omega h^2\,dx \int_\Omega u^2\,dx\right\}^{1/2} < \infty \tag{3.49}$$

One can also see that if $u(x) \in L_2$ and $l(x,x') \in L_2(\Omega \times \Omega)$, then $L(u(x)) \in L_2$.

3.4.2 CONVERGENCE AND MAPPING

In order to analyze the convergence of MC algorithms, the following equation is introduced as

$$u = \lambda L(u) + f \tag{3.50}$$

where λ denotes some parameters. $\lambda_1, \lambda_2, \ldots$ are the eigenvalues, where it is supposed that $|\lambda_1| \ge |\lambda_2| \ge \cdots$ Note that the matrices can be considered as linear operators. Define resolving operator R_λ by the equation,

$$I + \lambda R_\lambda = (I - \lambda L)^{-1} \tag{3.51}$$

where I is the identity operator.

MC algorithms are based on the representation

$$u = (I - \lambda L)^{-1} f = f + \lambda R_\lambda f \tag{3.52}$$

where $R_\lambda = L + \lambda L^2 + \cdots$

The systematic error of R_λ, where m terms are used, is calculated as

$$r_s = O[|\lambda/\lambda_1|^{m+1} m^{\rho-1}] \tag{3.53}$$

where ρ is the multiplicity of the root λ_1. When λ is approximately equal to λ_1, the sequence and the corresponding MC algorithm converge slowly. When $\lambda \ge \lambda_1$, the algorithm does not converge.

3.5 MARKOV CHAIN MC METHODS FOR THE EIGENVALUE PROBLEM

In the perspective of numerical point, the minimum in the eigen value matrix is more difficult than that of maximum calculation. However, it is more frequent to estimate the minimum by magnitude eigenvalue in physics and engineering, since the minimum is usually corresponding with the most stable state of the system.

3.5.1 FORMULATION OF THE EIGENVALUE PROBLEM

It is challenging to find efficient algorithms for the minimum eigenvalue evaluation, when the matrix is with a large size. Consider the following problem of evaluating eigenvalues $\lambda(A)$:

$$Ax = \lambda(A)x \tag{3.54}$$

It is assumed that
A is a given symmetric matrix, $a_{ij} = a_{ji}$ for all $i, j = 1, 2, \ldots, n;$.

$$\lambda_{\min} = \lambda_n < \lambda_{n-1} \le \lambda_{n-2} \le \cdots \le \lambda_2 < \lambda_1 = \lambda_{\max} \tag{3.55}$$

$$A = \left\{ a_{ij} \right\}_{i,j=1}^{n} = (a_1, \ldots, a_i, \ldots, a_n)^T \tag{3.56}$$

where $a_i = (a_{i1}, \ldots, a_{in})$ $i = 1, \ldots, n$ and the symbol T means transposition.
   The following norms of vectors

$$\|h\| = \|h\|_1 = \sum_{i=1}^{n} |h_i| \quad \|a_i\| = \|a_i\|_1 = \sum_{j=1}^{n} |a_{ij}| \tag{3.57}$$

and the matrices are used.

$$\|A\| = \|A\|_1 = \max_j \sum_{i=1}^{n} |a_{ij}| \tag{3.58}$$

In general, $\|A\| \ne \max_i \|a_i\|$
   By $\bar{A}$, we denote the matrix containing the absolute values of elements of a given matrix A:

$$\bar{A} = \left\{ |a_{ij}| \right\}_{i,j=1}^{n} \quad p_k(A) = \sum_{i=0}^{k} c_i A^i, c_i \in \mathbb{R} \tag{3.59}$$

where k represents the matrix polynomial degree.

As usual, $(v,h) = \sum_{i=1}^{n} v_i h_i$ denotes the inner product of vectors v and h.

The random variable ξ could be a randomly chosen component h_{α_k} of a given vector h. In this case, the meaning of $E(h_{\alpha_k})$ is the mathematical expectation of the value of a randomly chosen element of h.

$$D(\xi) = \sigma^2(\xi) = E(\xi^2) - [E(\xi)]^2 \qquad (3.60)$$

We denote the variance of the random variable ξ. In a special case of $p_k(A) = A^k$, the form $(v, p_k(A)h)$ becomes $(v, A^k h)$, $k \geq 1$.

Suppose that a real symmetric matrix A is diagonalizable,

$$x^{-1}Ax = diag(\lambda_1, \ldots, \lambda_n) \qquad (3.61)$$

If A is a symmetric matrix, then the values are real numbers, $\lambda \in \mathbb{R}$. The well-known power method gives an estimation for the dominant eigenvalue λ_1. This estimate uses the so-called Rayleigh quotient $\mu_k = \dfrac{(v, A^k h)}{(v, A^{k-1}h)}$.

where $v, h \in \mathbb{R}^n$ are arbitrary vectors. The Rayleigh quotient is used to obtain an approximation to λ_1.

$$\lambda_1 \approx \frac{(v, A^k h)}{(v, A^{k-1}h)} \qquad (3.62)$$

where k is an arbitrary large natural number.

To construct an algorithm for evaluating the eigenvalue of minimum modulus λ_n, one has to consider the following matrix of a polynomial.

$$p_i(A) = \sum_{k=0}^{i} q^k C_{m+k-1}^k A^k \qquad (3.63)$$

where C_{m+k-1}^k are binomial coefficients, and the characteristic parameter q is used as the acceleration parameter of the algorithm.

If $|q|\|A\| < 1$ and $i \to \infty$, then the polynomial becomes the re-solvent matrix

$$p_\infty(A) = p(A) = \sum_{k=0}^{\infty} q^k C_{m+k-1}^k A^k = [I - qA]^{-m} = R_q^m \qquad (3.64)$$

where $R_q = [I - qA]^{-1}$ is the re-solvent matrix of the equation.

$$x = qAx + h \qquad (3.65)$$

Values $q_1, q_2, \ldots$ for which the equation above is fulfilled are called characteristic values. The re-solvent operator,

$$R_q = [I - qA]^{-1} = I + A + qA^2 + \cdots \tag{3.66}$$

exists if the sequence converges. The systematic error of the presentation when m terms are used is

$$R_s = O\left[\left|q/q_1\right|^{m+1} m^{p-1}\right] \tag{3.67}$$

where ρ is the multiplicity of the root q_1. If $|q| < |q_1|$, the estimation accuracy of MC algorithm is satisfied When $|q| \geq |q_1|$, the algorithm does not converge for $q = q_* = 1$, but the solution of $x = qAx + h$ exists. In this case, one may apply a mapping of the spectral parameter q.

One can consider the ratio:

$$\lambda = \frac{(v, Ap(A)h)}{(v, p(A)h)} = \frac{(v, AR_q^m h)}{(v, R_q^m h)} \tag{3.68}$$

where $R_q^m h = \sum_{k=1}^{m} g_k^m c_k$ and g_k^m are computed. If $q < 0$, then

$$\frac{(v, AR_q^m h)}{(v, R_q^m h)} \approx \frac{1}{q}\left(1 - \frac{1}{\mu^k}\right) \approx \lambda_n \tag{3.69}$$

where $\lambda_n = \lambda_{\min}$ is the minimal by modulo eigenvalue, and μ^k is the approximation to the dominant eigenvalue of R_q.

If $|q| > 0$, then

$$\frac{(v, AR_q^m h)}{(v, R_q^m h)} \approx \lambda_1 \tag{3.70}$$

where $\lambda_1 = \lambda_{\max}$ is the dominant eigenvalue.

The approximate equations can be used to formulate efficient MC algorithms for evaluating both the dominant and the eigenvalue of minimum modulus of real symmetric matrices. Assume the set A, for calculating bilinear forms of matrix powers $(v, A^k h)$ with a probability error $R_{k,N}$ less than a given constant ε, and the probability $c < 1$ is also fixed. Obviously, for fixed ε and $c < 1$ the computational cost depends linearly on the number of iterations k and on the number of Markov chains N.

3.5.2 METHOD FOR CHOOSING THE NUMBER OF ITERATIONS K

To estimate the value of the bilinear form $(v, A^k h)$, so that with a given probability $P < 1$ the error is smaller than a given positive ε:

$$\left| (v, A^k h) - \frac{1}{N} \sum_{i=1}^{N} \theta_i^k \right| \leq \varepsilon \tag{3.71}$$

In the case of balanced errors,

$$R_{k,N} = R_{k,s} = \frac{\varepsilon}{2} \tag{3.72}$$

When a mapping procedure is applied, one may assume that there exists a positive constant $\alpha < 1$ such that $\alpha \geq \left| g_i^k \right| \times \|A\|$ for any i and k. Then,

$$\frac{\varepsilon}{2} \leq \frac{\left(\left| g_i^k \right| \|A\| \right)^{k+1} \|h\|}{1 - \left| g_i^k \right| \|A\|} \leq \frac{\alpha^{k+1} \|h\|}{1 - \alpha} \tag{3.73}$$

To choose k as the smallest natural number,

$$k \geq \frac{|\log \delta|}{|\log \alpha|} - 1 \quad \delta = \frac{\varepsilon(1 - \alpha)}{2\|h\|} \tag{3.74}$$

If a mapping procedure is not applied, the corresponding Neumann series converges fast enough, and then one assumes that a positive constant α, such that $\alpha \geq \|A\|$, exists.

3.5.3 METHOD FOR CHOOSING THE NUMBER OF CHAINS

To estimate the computational cost $\tau(A)$, we should estimate the number N of realizations of the random variable θ^k. If there exists a constant σ such that

$$\sigma \geq \sigma(\theta^k) \tag{3.75}$$

Then,

$$\varepsilon = 2R_N^k = 2c_p \sigma(\theta^k) N^{-0.5} \geq 2c_p \sigma N^{-0.5} \tag{3.76}$$

and

$$N \geq \left\{ \frac{2c_p \sigma}{\varepsilon} \right\}^2$$

Taking into account relations $k \geq \dfrac{|\log \delta|}{|\log \alpha|} - 1$, $\delta = \dfrac{\varepsilon(1 - \alpha)}{2\|h\|}$, and $N \geq \left\{ \dfrac{2c_p \sigma}{\varepsilon} \right\}^2$, the computational cost of biased MC algorithms can be evaluated.

REFERENCES

1. Grimmett, R., Stirzaker, D.R. *Probability and Random Processes*, 2nd Edition, Oxford: Clarendon Press. 1992. ISBN 0-19-853665-8.
2. Roshanian, J., Ebrahimi, M. Latin hypercube sampling applied to reliability-based multidisciplinary design optimization of a launch vehicle. *Aerospace Science and Technology*, **2013**, 28(1), 297–304.
3. Eugene, S., Donnell, O. *Incidence algebras, Pure and Applied Mathematics*, Marcel Dekker. 1997. ISBN 0-8247-0036-8.

4 Polynomial Chaos Expansion

4.1 FUNDAMENTAL DESCRIPTION OF PCE

The direct use of stochastic expansion of output responses and input random variables for representing uncertainty is an effective alternative. The stochastic expansion provides analytically appealing convergence properties based on the concept of a random process [1]. The polynomial chaos expansion (PCE) can reduce the computational effort of uncertainty quantification in engineering where system response is computed implicitly [2].

PCE is a probabilistic method consisting of the projection of the model output on the basis of orthogonal stochastic polynomials in the random inputs. The stochastic projection provides a compact and convenient representation of the model output.

The PCE stemmed from both Wiener and Ito's work on mathematical descriptions of irregularities [3]. Since Wiener introduced the concept of homogeneous chaos, the PCE has been successfully used for uncertainty analysis in various applications. A simple definition of the PCE for a Gaussian random response $u(\theta)$ as a convergent series is as follows:

$$u(\theta) = a_0 \Gamma_0 + \sum_{i=1}^{\infty} a_{i1} \Gamma_1(\xi_{i1}(\theta)) + \sum_{i_1=1}^{\infty} \sum_{i_2=1}^{i_1} a_{i_1 i_2} \Gamma_2(\xi_{i1}(\theta), \xi_{i2}(\theta)) + \cdots \tag{4.1}$$

where $\left\{\xi_i(\theta)\right\}_{i=1}^{\infty}$ is a set of Gaussian random variables; $\Gamma_p(\xi_{i1}, \dots, \xi_{ip})$ is the genetic element of a set of multidimensional Hermite polynomials, usually called homogeneous chaos of order p; $a_{i1}, \dots, a_{ip}$ are deterministic constants; and θ represents an outcome in the space of possible outcomes of a random event.

PCE is convergent in the mean-square sense and the p th order PCE consists of all orthogonal polynomials of order p, including any combinations of $\left\{\xi_i(\theta)\right\}_{i=1}^{\infty}$; furthermore, $\Gamma_p \perp \Gamma_q$ for $p \neq q$. This orthogonality greatly simplifies the procedure of statistical calculations, such as moments. Therefore, PCE can be used to approximate non-Gaussian distributions using a least-squares scheme.

The general expression to obtain the multidimensional Hermite polynomials is given by

$$\Gamma_p(\xi_{i1}, \dots, \xi_{ip}) = (-1)^n \frac{\partial^n e^{\frac{1}{2}\vec{\xi}^T \vec{\xi}}}{\partial \xi_{i1}, \dots, \partial \xi_{ip}} e^{\frac{1}{2}\vec{\xi}^T \vec{\xi}} \tag{4.2}$$

where the vector $\vec{\xi}$ consists of n Gaussian random variables.

DOI: 10.1201/9781003226628-5

Then $u(\theta)$ can be written more simply as

$$u(\theta) = \sum_{i=0}^{P} b_i \Psi_i(\vec{\xi}(\theta)) \tag{4.3}$$

where b_i and $\Psi_i(\vec{\xi}(\theta))$ are one-to-one correspondences between the coefficients $a_{i1},\ldots,a_{ip}$ and the functions $(\xi_{i1},\ldots,\xi_{ip})$, respectively.

4.2 STOCHASTIC APPROXIMATION

One of the important applications of stochastic expansion is the nonintrusive formulation to create a surrogate model. If we fit curvilinear data, the following regression model can be considered:

$$Y(x) = \beta_0 F_0(x) + \beta_1 F_1(x) + \beta_2 F_2(x) + \beta_3 F_3(x) \tag{4.4}$$

where β_0, β_1, β_2, and β_3 represent the mean, linear, quadratic, and cubic effects, respectively, of the response; Y and $F_i(x)$ are basis polynomials.

The use of orthogonal polynomial can eliminate collinearity and ill-conditioned problems. The basic idea of the stochastic approximation utilizing stochastic expansion is to select an appropriate basis function to represent the response of uncertain systems. The PCE, which employs orthogonal basis functions and is mean-square convergent, is a good choice for estimating the response variability of uncertain systems.

PCE can be used to represent the response of an uncertain system in the nonintrusive formulation. The basic idea of this approach is to project the response and stochastic system operator onto the stochastic space spanned by PCE, with the projection coefficients, b_i, being evaluated through an efficient sampling scheme. We first define the vector x at a particular point $(\xi_i, \cdots, \xi_m)$ of random variables.

The estimated response at this point is

$$y(x) = x^T \hat{\beta} \tag{4.5}$$

where $\hat{\beta}$ is a set of undetermined coefficients of PCE.

Generally, the method of least squares is used to obtain the regression coefficients for n sample values of x and y as

$$\hat{\beta} = (X^T X)^{-1} X^T Y \tag{4.6}$$

where X is a $n*p$ matrix of the levels of the regression variables and Y is a $n*l$ vector of the responses.

The fitted model $\hat{Y}$ and the residuals e are

$$\hat{Y} = X\hat{\beta} \tag{4.7}$$

and $e = Y - \hat{Y}$

TABLE 4.1

Representation of Various Distributions as Functions of Normal Random Variables

Distribution Type	Transformation
Normal (μ, σ)	$\mu + \sigma \xi$
Lognormal (μ, σ)	$\exp(\mu + \sigma \xi)$
Uniform (a, b)	$a + (b-a)\left(\dfrac{1}{2} + \dfrac{1}{2}\mathrm{erf}(\dfrac{\xi}{\sqrt{2}})\right)$
Exponential (λ)	$-\dfrac{1}{\lambda}\log\left(\dfrac{1}{2} + \dfrac{1}{2}\mathrm{erf}(\dfrac{\xi}{\sqrt{2}})\right)$
Gamma (a, b)	$ab\left(\xi\sqrt{\dfrac{1}{9a}} + 1 - \dfrac{1}{9a}\right)^{3}$

A confidence interval indicates a range of values that likely contains the analysis results. Generally, the confidence interval of each parameter includes two parts: the confidence level and the margin of error.

The confidence level expresses the probability with which the interval contains the true parameter value. The margin of error represents how accurate our approximation of the true parameter value is. When x_0 is the vector at a particular point $(\xi_i, \cdots, \xi_m)$ of random variables, then the estimated mean response at that particular point is

$$\hat{y}(x_0) = x_0^T \hat{\beta} \tag{4.8}$$

where $\hat{\beta}$ is a set of undetermined coefficients of PCE. A $100(1-\alpha)$ percent confidence interval at the particular point x_0 is

$$\hat{y}(x_0) - t_{\frac{\alpha}{2}, v}\sqrt{\sigma^2 x_0^T (X^T X)^{-1} x_0} \leq \mu \leq \hat{y}(x_0) + t_{\frac{\alpha}{2}, v}\sqrt{\sigma^2 x_0^T (X^T X)^{-1} x_0} \tag{4.9}$$

where σ^2 is the variance, v is the degree of freedom, and α indicates the $100(1-\alpha)$ th percentile of the t distribution. The point x_0 is not limited to one of the sampling points used, since the interval includes the results of random samples from the given population with mean μ.

Gamma distribution or exponential distributions should be represented by a normal probability distribution (Table 4.1).

4.3   HERMITE POLYNOMIALS AND GRAM–CHARLIER SERIES

Before beginning the topic of the KL transform, it is useful to know several properties of the Hermite polynomial, which is the basis of the PCE. The construction of Hermite polynomials was described by Pafnuty Chebyshev and Charles Hermite. The second-order differential equation is given by

$$\frac{d^2y}{dx^2} - x\frac{dy}{dx} + ny = 0 \quad \text{or} \quad \frac{d^2y}{dx^2} - 2x\frac{dy}{dx} + 2ny = 0 \tag{4.10}$$

where n is a positive integer. The corresponding possible solutions are

$$H_n(x) = (-1)^n e^{x^2/2} \frac{d^n e^{-x^2/2}}{dx^n}$$

or

$$H_n(x) = (-1)^n e^{x^2} \frac{d^n e^{-x^2}}{dx^n} \tag{4.11}$$

These polynomials are called the Hermite polynomials. Although these two equations are not equivalent, the first is a linear rescaling of the domain of the second. Since the n th derivative of the normal density function $\varphi(x) = 1/\sqrt{2\pi}\, e^{-x^2/2}$ is included in this equation, the definition of the equation is often used in probabilistic analysis.

The orthogonal properties of the Hermite polynomials are given in the interval $[-\infty, +\infty]$ with respect to the weight function of $e^{-x^2/2}$ or e^{-x^2}:

$$\int_{-\infty}^{\infty} e^{-x^2/2} H_n(x) H_m(x) dx = n! \sqrt{2\pi} \delta_{nm} \tag{4.12}$$

and

$$\int_{-\infty}^{\infty} e^{-x^2} H_n(x) H_m(x) dx = 2^n n! \sqrt{\pi} \delta_{nm}$$

It implies that the Hermite polynomials are orthogonal with respect to the Gaussian distribution. Also, note that the weight functions, $e^{-x^2/2}$ or e^{-x^2}, help keep the integral from reaching infinity over the interval from $-\infty$ to ∞, since the exponential functions converge to zero much faster than the polynomials when x is large.

$$H_n(x) = (-1)^n e^{x^2/2} \frac{d^n}{dx^n} e^{-x^2/2} \tag{4.13}$$

$$\begin{cases} H_0(x) = 1 \\ H_1(x) = x \\ H_2(x) = x^2 - 1 \\ H_3(x) = x^3 - 3x \\ H_4(x) = x^4 - 6x^2 + 3 \\ H_5(x) = x^5 - 10x^3 + 15x \end{cases} \tag{4.14}$$

When this orthogonal property of the Hermite polynomials is used to estimate the probability density function, the procedure is known as the Gram–Charlier method. The basic idea of the Gram–Charlier method is that the density function of the Gaussian distribution and its derivatives provide a series expansion to represent an arbitrary density function. The Gram–Charlier series is given by

$$f(x) = b_0 \varphi(x) + b_1 \varphi'(x) + b_2 \varphi''(x) + \cdots \tag{4.15}$$

where $f(x)$ is the unknown probability density function, and $\varphi''(x)$ is the n th derivative of the normal density function, $\varphi(x) = 1/\sqrt{2\pi}\, e^{-x^2/2}$

$$H_n(x) = (-1)^n \frac{\varphi^{(n)}(x)}{\varphi(x)} \tag{4.16}$$

$$\varphi^{(n)}(x) = (-1)^n \varphi(x) H_n(x) \tag{4.17}$$

Then,

$$f(x) = \varphi(x)\left[b_0 H_0(x) - b_1 H_1(x) + b_2 H_2(x) + \cdots \right]$$

$$= \varphi(x) \sum_{m=0}^{\infty} (-1)^m b_m H_m(x) \tag{4.18}$$

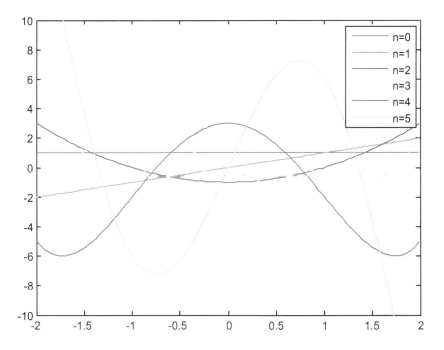

FIGURE 4.1   Hermite polynomials.

To find the b_i coefficient, multiply both sides by $H_n(x)$ and integrate from $-\infty$ to ∞. The result is

$$\int_{-\infty}^{\infty} f(x)H_n(x)dx = \sum_{m=0}^{\infty} (-1)^m b_m \int_{-\infty}^{\infty} \varphi(x)H_n(x)H_m(x)dx \tag{4.19}$$

Because of the orthogonal property of the Hermite polynomials

$$\int_{-\infty}^{\infty} \varphi(x)H_n(x)H_m(x)dx = \begin{cases} n! & \text{for } n = m \\ 0 & \text{for } n \neq m \end{cases} \tag{4.20}$$

This property can be used to compute the coefficient,

$$b_n = \frac{(-1)^n}{n!} \int_{-\infty}^{\infty} f(x)H_n(x)dx \tag{4.21}$$

4.4   KARHUNEN–LOEVE TRANSFORM

In stochastic analysis, the effective representation of uncertainty information and appropriate application of the information to evaluate the safety are the primary challenges. However, the uncertainty space is extended and coupled in the space and time domains. The description of space and time variation can be expressed by the concept of the random fields. KL transform is an efficient method to solve the spatial correlation and perform the dimension reduction for the random variables.

Due to the simplicity of its procedure, the most widely used method of multivariate data analysis is the orthogonal transform method. The KL transform is a feasible tool with multiple uses for uncertainty analysis.

The KL expansion can be viewed as part of a general orthogonal series expansion. Consider a general series expansion of $f(x)$ with a complete set of orthogonal and normalized base functions $\phi_i(x)$:

$$f(x) = \sum_{i=1}^{N} b_i \phi_i(x) \tag{4.22}$$

where the coefficients b_i represent the projection of $f(x)$ on the basis function $\phi_i(x)$, and b_i is obtained by

$$b_i = \int f(x)\phi_i(x)\,dx \tag{4.23}$$

The condition of uncorrelated coefficients yields

$$\left\langle (b_i - \mu_i)(b_j - \mu_j) \right\rangle = \lambda_j \delta_{ij} \tag{4.24}$$

where $\langle \bullet \rangle$ indicates the expected value operation, δ_{ij} is the Kronecker delta, and μ is the mean of the coefficients b. This restriction results in the following eigenvalue analysis of the covariance function

$$\lambda_i \phi_i(x) = \int K(x,y)\phi_i(y)\,dy \qquad (4.25)$$

where $\phi_i(x)$ and λ_i are the eigenfunctions and eigenvalues of the covariance function $K(x,y)$, respectively, and x and y are the temporal or spatial coordinates:

$$K(x,y) = \langle (b(x) - \mu(x))(b(y) - \mu(y)) \rangle \qquad (4.26)$$

where $\mu(x)$ is the mean of the coefficients $b(x)$.

The series of the eigenfunctions and the eigenvalues form the KL expansion:

$$w(x) = \sum_{i=1}^{\infty} \sqrt{\lambda_i}\,\xi_i \phi_i(x) \qquad (4.27)$$

where ξ_i is a set of uncorrelated random variables, and this expansion expresses the projection of the random process $w(x)$.

In the discrete case,

$$[P][\Lambda] = [K][P] \qquad (4.28)$$

where the covariance matrix $[K]$ is a symmetric and nonnegative definite matrix, and $[P]$ and $[\Lambda]$ are the orthogonal eigenvector matrix and the eigenvalue matrix, respectively.

Consequently, the orthogonal decomposition of the covariance matrix provides the product of the matrices of eigenvectors and eigenvalues:

$$[K] = [P][\Lambda][P]^T \qquad (4.29)$$

or

$$[K] = [A][A]^T$$

where $[A]$ is the transform matrix chosen as $[A] = [P][\Lambda]^{1/2}$.

The transform matrix $[A]$ can be employed to yield the correlated random vector T:

$$[T] = [A][X] \qquad (4.30)$$

where $[X]$ is the $(n*1)$ matrix of uncorrelated random variables X_j, $(j = 1,\ldots,n)$, and the transformed matrix, $[T]$, possesses a given covariance matrix $[K]$.

In addition to generating the dependent random variables, T, the KL transform can be used to reduce the dimension of the random variables. The main advantage of this procedure is to permit a significant reduction in the number of uncorrelated random variables that represent random fields, especially for high levels of correlation.

4.5 KARHUNEN–LOEVE EXPANSION

The KL expansion can be derived based on the analytical properties of its covariance function. Let the covariance function be specified by the exponential covariance with a variance of C_0, a correlation length of $1/h$, and two different locations of x_1 and x_2 defined in $\left[-a \le x_1, x_2 \le a\right]$:

$$K(x_1, x_2) = C_0 e^{-h|x_1 - x_2|}, \quad -a \le x_1, x_2 \le a \tag{4.31}$$

Then $\lambda_i \phi_i(x) = \int K(x, y)\phi_i(y)dy$ can be written as

$$\lambda \phi(x_1) = \int_{-a}^{a} C_0 e^{-h|x_1 - x_2|}\phi(x_2)\,dx_2 \tag{4.32}$$

We need to solve the above equation by converting the integral equation to a differential equation and then substituting the solution back into the integral equation. To eliminate the absolute magnitude sign,

$$\lambda \phi(x_1) = \int_{-a}^{x_1} C_0 e^{-h(x_1 - x_2)}\phi(x_2)\,dx_2 + \int_{x_1}^{a} C_0 e^{h(x_1 - x_2)}\phi(x_2)\,dx_2 \tag{4.33}$$

Defining $\omega = (2C_0 h - h^2 \lambda)/\lambda$

$$\phi''(x_1) + \omega^2 \phi(x_1) = 0 \quad -a \le x_1 \le a \tag{4.34}$$

Let $x_1 = t$

$$\phi(t) = c_1 e^{j\omega t} + c_2 e^{j\omega t}, \quad \omega^2 \ge 0 \tag{4.35}$$

where c_1 and c_2 are constants.

Applying the boundary condition yields

$$(h - \omega \tan(\omega a))(\omega + h \tan(\omega a)) = 0$$

$$h - \omega \tan(\omega a) = 0 \quad \text{or} \quad \omega + h \tan(\omega a) = 0$$

The values of ω can be determined graphically or numerically, and the corresponding eigenvalues are

$$\lambda_i = \frac{2C_0 h}{h^2 + \omega_i^2}, \quad i = 1, 2, 3, \ldots, n \tag{4.36}$$

The resulting eigenfunction is

$$\phi_i(t) = \frac{\cos \omega_i t}{\sqrt{a + \dfrac{\sin(2\omega_i a)}{2\omega_i}}} \quad \text{(for, } i = \text{odd)}, \quad -a \le t \le a \tag{4.37}$$

$$\phi_i(t) = \frac{\sin \omega_i t}{\sqrt{a - \dfrac{\sin(2\omega_i a)}{2\omega_i}}} \quad \text{(for, } i = \text{even)}, \quad -a \le t \le a \tag{4.38}$$

After the graphical or numerical solution of transcendental equations for ω_i, the eigenfunctions can be given as a set of periodic sines and cosines at approximately $(i-1)\pi/2$.

4.6 COMPARISON AND DISCUSSION

PCE is used to represent stochastic responses, and the KL expansion is used to represent the input of random fields in the intrusive formulation procedure. This method is also known as the spectral stochastic finite element method (SSFEM) and yields appropriate results for a wide range of random fluctuations [4].

Recall that in the KL expansion, a series of eigenfunctions and eigenvalues with a set of random variables ξ_i represent the random process. The eigenvalues and eigenfunctions can be obtained. Let $w(x,\theta)$ denote a random process, so that the function can be expanded in the following form, truncated to M terms:

$$w(x,\theta) = \overline{w}(x) + \sum_{i=1}^{M} \sqrt{\lambda_i}\, \xi_i(\theta)\phi_i(x) \tag{4.39}$$

where $\overline{w}(x)$ denotes the expected value of the random process and θ represents an outcome in the space of possible outcomes of a random event.

Suppose Young's modulus is a Gaussian random field. Then, the elasticity matrix D can be written as

$$D(x,\theta) = w(x,\theta)D_0 \tag{4.40}$$

where D_0 is a constant matrix similar to the one in deterministic finite element analysis.

The element stiffness matrix is

$$K^e(\theta) = K_0^{(e)} + \sum_{i=1}^{M} K_i^{(e)}\xi_i(\theta) \tag{4.41}$$

where $K_0^{(e)}$ is the mean element stiffness matrix and

$$K_i^{(e)} = \sqrt{\lambda_i} \int_{\Omega_e} \phi(x) B_e^T D_0 B_e d\Omega^e \tag{4.42}$$

where B_e is the matrix determined from the shape functions and geometric condition of the finite element.

Assembling the above element contributions in the finite element analysis procedure eventually gives

$$\left[K_0 + \sum_{i=1}^{M} K_i \xi_i(\theta) \right] u(\theta) = f \tag{4.43}$$

$$K_0 \left[I + \sum_{i=1}^{M} K_0^{-1} K_i \xi_i(\theta) \right] u(\theta) = f \tag{4.44}$$

$$\left[I + \sum_{i=1}^{M} K_0^{-1} K_i \xi_i(\theta) \right] u(\theta) = u_0 = K_0^{-1} f \tag{4.45}$$

It leads to

$$u(\theta) = \left[I + \sum_{i=1}^{M} K_0^{-1} K_i \xi_i(\theta) \right]^{-1} u_0 \tag{4.46}$$

Now, the displacement vector can be obtained by the Neumann series

$$u(\theta) = \sum_{i=0}^{\infty} (-1)^i \left(\sum_{n=1}^{M} K_0^{-1} K_n \xi_n(\theta) \right)^i u_0 \tag{4.47}$$

Applying the expected value operator, the mean of the response yields

$$E[u] = \sum_{i=0}^{\infty} (-1)^i E\left[\left(\sum_{n=1}^{M} K_0^{-1} K_n \xi_n(\theta) \right)^i u_0 \right] \tag{4.48}$$

In a general case, the covariance matrix yields

$$Cov[u,u] = \sum_{i=0}^{\infty} \sum_{j=0}^{\infty} (-1)^{i+j} E\left[\left(\sum_{n=1}^{M} K_0^{-1} K_n \xi_n \right)^i K_0^{-1} f \times f^T K_0^{-T} \left(\sum_{m=1}^{M} K_m^T K_0^{-T} \xi_m \right)^j \right] \tag{4.49}$$

The KL expansion requires known covariance functions to obtain the eigenvalues and eigenfunctions. Since the covariance function of stochastic responses often is not known, PCE is used to represent stochastic responses in SSFEM.

Recalling the definition of PCE, $u(\theta)$ can be projected on the expansion

$$u(\theta) = \sum_{j=0}^{\infty} b_j \psi_j(\theta) \tag{4.50}$$

Then in the finite element model,

$$\left(\sum_{i=1}^{\infty} K_i \xi_i(\theta) \right) \left(\sum_{j=0}^{\infty} b_j \psi_j(\theta) \right) = f \tag{4.51}$$

Truncating the KL expansion after M terms and PCE after P terms results in

$$\sum_{i=0}^{M} \sum_{j=0}^{P} \xi_i(\theta) \psi_j(\theta) K_i b_j - f = \varepsilon \tag{4.52}$$

Minimization of the residual leads to an accurate approximation of the solution $u(\theta)$. This requires the residual to be orthogonal to the approximating space spanned by the PCE. Orthogonality requires the inner product to be equal to zero, namely,

$$E[\varepsilon \cdot \Psi_k] = 0 \tag{4.53}$$

Thus, the expected value of $\sum_{i=0}^{M} \sum_{j=0}^{P} \xi_i(\theta) \psi_j(\theta) K_i b_j - f = \varepsilon$ becomes

$$\sum_{i=0}^{M} \sum_{j=0}^{P} E\left[\xi_i(\theta) \Psi_j(\theta) \Psi_k(\theta) \right] K_i b_j = E\left[f \Psi_k(\theta) \right] \quad k = 0, \ldots, P \tag{4.54}$$

which can be rewritten as

$$\sum_{j=0}^{P} K_{jk} b_j = f_k \tag{4.55}$$

$$K_{jk} = \sum_{i=0}^{M} C_{ijk} K_i \tag{4.56}$$

$$C_{ijk} = E\left[\xi_i(\theta) \Psi_j(\theta) \Psi_k(\theta) \right] \tag{4.57}$$

$$f_k = E\left[f \Psi_k(\theta) \right] \tag{4.58}$$

In the matrix, we can rewrite this as

$$
\begin{bmatrix}
K^{(0,0)} & K^{(0,1)} & \cdots & K^{(0,P)} \\
K^{(1,0)} & K^{(1,1)} & \cdots & K^{(1,P)} \\
\vdots & \vdots & \ddots & \vdots \\
K^{(P,0)} & K^{(P,1)} & \cdots & K^{(P,P)}
\end{bmatrix}
\begin{Bmatrix}
b^{(0)} \\
b^{(1)} \\
\vdots \\
b^{(P)}
\end{Bmatrix}
=
\begin{Bmatrix}
f^{(0)} \\
f^{(1)} \\
\vdots \\
f^{(P)}
\end{Bmatrix}
\tag{4.59}
$$

There is a $P+1$ dimensional matrix.

Once the system is computed with the coefficient vectors b_j, the statistics of the solution can be readily obtained. The mean and covariance matrix of $u(\theta)$ can be obtained as

$$
E[u(\theta)] = b_0
\tag{4.60}
$$

$$
Cov[u,u] = E\left[(u - u_0)(u - u_0)^T\right] = \sum_{i=1}^{P} E\left[\Psi_j^2\right] b_j b_j^T
\tag{4.61}
$$

Multidimensional Hermite orthogonal polynomials are firstly proposed to represent the Gaussian stochastic process by Wiener, based on which a SSFEM is developed by Ghanem and Spanos and widely used in various applications, including structural mechanics, fluid flow, and so on [5].

The efficient method for uncertainty analysis (UA) aims to reduce the time for a single reliability analysis or moment evaluation procedure, and the advanced formulation is to reduce the number of UA. Establish an explicit relation between the probability of failure/moments and the design variables.

- The sequential quadratic programming (SQP) method is one of the most used methods, a standard mathematical programming algorithm for solving nonlinear programming optimization problems. This method can assure a local optimum but not a global one. This shortcoming may be avoided by multiple initial designs (evolutionary algorithm and genetic algorithm for which no gradient information is needed).
- The perturbation method is based on the Taylor series expansion in terms of a set of zero-mean random variables. It can be used advantageously in cases where the random fluctuations are small compared with the nominal structure, such that terms of order two or higher are negligible. The perturbation method can determine the uncertainties without large dispersion, especially for moment evaluations of the random response. There are fewer applications of such methods to reliability analysis.
- In the framework of the PCE method, the random response can be approximated with acceptable accuracy.

The main advantage of the PCE compared to the KL expansion is that the covariance structure is not required.

PCE, from the efficiency point of view, is more applicable for problems with a small number of random inputs. This situation is more involved with static problems rather than dynamic ones since the stochastic excitation is discretized by an uncertainty sequence with a high dimension.

REFERENCES

1. Crestauxa, T., Maitreb, O., Martinezc, J.M. Polynomial chaos expansion for sensitivity analysis. *Reliability Engineering & System Safety*, **2009**, 94(7), 1161–1172.
2. Lia, R., Ghanem, R. Adaptive polynomial chaos expansions applied to statistics of extremes in nonlinear random vibration. *Probabilistic Engineering Mechanics*, **1998**, 13(2), 125–136.
3. Wan, X.L., Karniadakis, G.E. Multi-element generalized polynomial chaos for arbitrary probability measures. *Journal of Scientific Computing*, **2006**, 28(3), 901–928.
4. Shinozuka, M., Deodatis, G. Simulation of stochastic processes by spectral representation. *ASME, Applied Mechanics Review*, **1991**, 44(4), 191–204.
5. Dham, S, Ghanem, R. Finite element analysis of multiphase flow in porous media with the polynomial chaos expansion, *Second International Conference on Computational Stochastic Mechanics*, Athens, Greece (June 1994).

5 Stochastic Finite Element Method

5.1 METHODS FOR DISCRETIZATION OF RANDOM FIELDS

From a mathematical point of view, a stochastic finite element method (SFEM) is a useful approach for the solution of stochastic partial differential equations (PDEs) with satisfied convergence and accuracy. The considerable attention that the SFEM has received over the last decade can be mainly attributed to the spectacular growth of computational power making the efficient treatment of large-scale problems possible [1]. The SFEM is an extension of the classical deterministic finite element approach with broad application in mechanic computation, material property analysis, reliability evaluation in engineering, and so on. The uncertainty in material property, stochastically distributed defects, and the instability of microstructure boundaries are challenging issues in the framework of a deterministic finite element method.

A fundamental issue in the SFEM is the modeling of the uncertainty characterizing the system parameters (input). This uncertainty is quantified by using the theory of stochastic functions (processes/ fields). The first step in the analysis of uncertain systems (in the framework of SFEM) is the representation of the input of the system. This input usually consists of the mechanical and geometric properties as well as of the loading of the system. Characteristic examples are the Young modulus, Poisson ratio, yield stress, cross-section geometry of physical systems, material and geometric imperfections of shells, earthquake loading, wind loads, waves, and so on. A convenient way for describing these uncertain quantities in time and/or space has always been the implementation of stochastic processes and fields, the probability distribution, and the correlation structure which can be defined through experimental measurements.

However, in most cases, due to the lack of relevant experimental data, assumptions are made regarding these probabilistic characteristics. Two main categories of stochastic processes and fields can be defined based on their probability distribution: Gaussian and non-Gaussian. From the wide variety of methods developed for the simulation of Gaussian stochastic processes and fields, two are most often used in applications: the spectral representation method and the Karhunen–Loève (KL) expansion.

5.1.1 Point Discretization Methods

5.1.1.1 The Midpoint Method

In the context of the finite element method, a spatial discretization of the system geometry is utilized for the approximation of the mechanical response of the structure.

DOI: 10.1201/9781003226628-6

This method consists in approximating the random field in each element Ω_e by a single random variable defined as the value of the field at the centroid x_c of this element:

$$\hat{H}(x) = H(x_c), \quad x \in \Omega_e \tag{5.1}$$

The approximated field $\hat{H}(\cdot)$ is then entirely defined by the random vector $\chi = \left\{ H(x_c^1), \ldots, H(x_c^{N_e}) \right\}$ (with N_e being the number of elements in the mesh). Its mean μ and covariance matrix $\sum_{\chi\chi}$ are obtained from the mean, variance, and autocorrelation coefficient functions of $H(\cdot)$ evaluated at the element centroids. Each realization of $\hat{H}(\cdot)$ is piecewise constant, with the discontinuities being localized at the element boundaries.

5.1.1.2   The Shape Function Method

This method approximates $H(\cdot)$ in each element using nodal values x_i and shape functions as follows,

$$\hat{H}(x) = \sum_{i=1}^{q} N_i(x) H(x_i), \quad x \in \Omega_e \tag{5.2}$$

where q is the number of nodes of an element, $e_i x_i$ represents the coordinates of the i-th node, and N_i represents polynomial shape functions associated with the element. The approximated field $\hat{H}(\cdot)$ is obtained in this case from $\chi = \left\{ H(x_1), \ldots, H(x_N) \right\}$, where $\left\{ x_i, i = 1, \ldots, N \right\}$ is the set of the nodal coordinates of the mesh.

The mean value and covariance of the approximated field $\hat{H}(\cdot)$ read:

$$E\left[\hat{H}(x) \right] = \sum_{i=1}^{q} N_i(x) \mu(x_i) \tag{5.3}$$

$$Cov\left[\hat{H}(x), \hat{H}(x') \right] = \sum_{i=1}^{q} \sum_{j=1}^{q} N_i(x) N_j(x') Cov\left[H(x_i), H(x_j) \right] \tag{5.4}$$

Each realization of $\hat{H}(\cdot)$ is a continuous function over Ω, which is an advantage over the midpoint method.

5.1.2   Average Discretization Methods

5.1.2.1   Spatial Average

Provided a mesh of the structure is available, it defines the approximated field in each element as a constant being computed as the average of the original field over the element,

$$\hat{H}(x) = \frac{\int_{\Omega_e} H(x) d\Omega_e}{|\Omega_e|} \equiv \bar{H}_e, \quad x \in \Omega_e \tag{5.5}$$

Vector χ is then defined as the collection of these random variables, that is $\chi^T = \{\bar{H}_e, e = 1,\dots,N_e\}$. The mean and covariance matrix of χ are computed from the mean and covariance function of $H(x)$ as integrals over the domain Ω_e.

However, there are difficulties involved in this method, and the approximation for nonrectangular elements may lead to a nonpositive definite covariance matrix. The probability density function of each random variable χ_i is almost impossible to obtain except for Gaussian random fields.

5.1.2.2 The Weighted Integral Method

In the context of linear elasticity, the main idea is to consider the element stiffness matrices as basic random quantities. More precisely, using standard finite element notations, the stiffness matrix associated with a given element occupying a volume Ω_e reads

$$k^e = \int_{\Omega_e} B^T \cdot D \cdot B d\Omega_e \tag{5.6}$$

where D denotes the elasticity matrix and B is a matrix that relates the components of strains to the nodal displacements.

Consider the elasticity matrix obtained as a product of a deterministic matrix by a univariate random field,

$$D(x,\theta) = D_0 [1 + H(x,\theta)] \tag{5.7}$$

where D_0 is the mean value and $H(x,\theta)$ is a zero-mean process. Thus, the above equation can be rewritten as

$$k^e(\theta) = k_0^e + \Delta k^e(\theta), \quad \Delta k^e(\theta) = \int_{\Omega_e} H(x,\theta) B^T \cdot D_0 \cdot B d\Omega_e \tag{5.8}$$

Furthermore, the elements in the matrix B are obtained by derivation of the element shape functions with respect to the coordinates. Hence, they are polynomials in the latter. A given member of Δk^e is thus obtained after matrix product as

$$\Delta k_{ij}^e(\theta) = \int_{\Omega_e} P_{ij}(x,y,z) H(x,\theta) d\Omega_e \tag{5.9}$$

where the coefficients of the polynomial P_{ij} are obtained from those of B and D_0. Then,

$$P_{ij}(x,y,z) = \sum_{l=1}^{N} a_{ij}^l x^{\alpha_l} y^{\beta_l} z^{\gamma_l} \tag{5.10}$$

where N is the number of monomials in P_{ij}, each of them corresponding to a set of exponents $\{\alpha_l, \beta_l, \gamma_l\}$. Then,

$$\chi_i^e(\theta) = \int_{\Omega_e} x^{\alpha_l} y^{\beta_l} z^{\gamma_l} H(x,\theta) d\Omega_e \tag{5.11}$$

It follows that

$$\Delta k_{ij}^e(\theta) = \sum_{l=1}^{N} a_{ij}^l \chi_i^e(\theta) \tag{5.12}$$

Collecting now the coefficients a_{ij}^l in a matrix Δk_i^e, the stochastic element stiffness matrix can finally be written as

$$k^e = k_0^e + \sum_{l=1}^{N} \Delta k_i^e \chi_i^e \tag{5.13}$$

In the above equations, k_0^e and Δk_i^e are the deterministic matrices and χ_i^e represents random variables.

The original random field is actually projected onto the space of polynomials involved in the B-matrices, which is basically onto the space spanned by the shape functions of the finite elements. This is an implicit kind of discretization similar to the shape function approach. Moreover, if the correlation length of the random field is small compared to the size of the integration domain Ω_e, the accuracy of the method is questionable. Indeed, the shape functions usually employed for elements with constant properties may not give good results when these properties are rapidly varying in the element. The problem of accuracy of the weighted integrals approach seems not to have been addressed in detail in the literature.

5.1.3 Series Expansion Methods

The discretization methods presented up to now involved a finite number of random variables having a straightforward interpretation: point values or local averages of the original field. In all cases, these random variables can be expressed as weighted integrals of $H(\cdot)$ over the volume of the system,

$$\chi_i(\theta) = \int_{\Omega} H(x,\theta) w(x) d\Omega \tag{5.14}$$

The weight functions ($w(x)$) are summarized in Table 5.1. In the table, $\delta(\cdot)$ denotes the Dirac function and 1_{Ω_e} is the characteristic function of the element e defined by

$$1_{\Omega_e}(x) = \begin{cases} 1 & \text{if } x \in \Omega_e \\ 0 & \text{otherwise} \end{cases} \tag{5.15}$$

TABLE 5.1
Weight Functions and Deterministic Basis

Method	Weight Function $w(x)$	$\varphi_i(x)$
MP	$\delta(x - x_c)$	$1_{\Omega_e}(x)$
SA	$\dfrac{1_{\Omega_e}(x)}{\lvert\Omega_e\rvert}$	$1_{\Omega_e}(x)$
SF	$\delta(x - x_i)$	Polynomial shape functions $N_i(x)$
OLE	$\delta(x - x_i)$	

By means of these random variables $\chi_i(\theta)$, the approximated field can be expressed as a finite summation.

$$\hat{H}(x,\theta) = \sum_{l=1}^{N} \chi_i(\theta)\varphi_i(x) \tag{5.16}$$

The deterministic functions $\varphi_i(x)$ are reported in the table.

The discretization methods presented in the present section aim at expanding any realization of the original random field $H(x,\theta_0) \in L^2(\Omega)$ over a complete set of deterministic functions. The discretization occurs thereafter by truncating the obtained series after a finite number of terms.

The KL expansion of a random field $H(\cdot)$ is based on the spectral decomposition of its autocovariance function $C_{HH}(x,x') = \sigma(x)\sigma(x')\rho(x,x')$. The set of deterministic functions over which any realization of the field $H(x,\theta_0)$ is expanded is defined by the eigenvalue problem,

$$\forall\, i = 1,\ldots \int_\Omega C_{HH}(x,x')\varphi_i(x')\,d\Omega_{x'} = \lambda_i\varphi_i(x) \tag{5.17}$$

It is a Fredholm integral equation, the kernel $C_{HH}(\cdot,\cdot)$ being an autocovariance function; it is bounded, symmetric, and positive definite. Thus, the set of $\{\varphi_i\}$ forms a complete orthogonal basis of $L^2(\Omega)$. The set of eigenvalues (spectrum) is moreover real, positive, numerable, and has zero as the only possible accumulation point. Any realization of $H(\cdot)$ can thus be expanded over this basis as follows:

$$H(x,\theta) = \mu(x) + \sum_{i=1}^{\infty} \sqrt{\lambda_i}\xi_i(\theta)\varphi_i(x) \tag{5.18}$$

where $\xi_i(\theta)$ denotes the coordinates of the realization of the random field with respect to the set of deterministic functions $\{\varphi_i\}$. Taking now into account all possible realizations of the field, $\{\xi_i\}$ becomes a numerable set of random variables.

5.2   PERTURBATION METHOD

The probability theory was introduced in mechanics to estimate the response variability of a system, which is the dispersion of the response around a mean value when the input parameters themselves vary around their means. The aim is to understand how uncertainties in the input propagate through the mechanical system. For this purpose, second-order statistics of the response are to be evaluated.

Suppose the input randomness in geometry, material properties, and loads is described by a set of N random variables, each of them being represented as the sum of its mean value and a zero-mean random variable α_i. The input variations around the mean are thus collected in a zero-mean random vector $\alpha = \{\alpha_1, \ldots \alpha_N\}$. In the context of finite element analysis, the second-moment methods aim at evaluating the statistics of the nodal displacements, strains, and stresses from the mean values of the input variables and the covariance matrix of α.

The perturbation method introduced in the late 1970s has been employed in a large number of studies. It uses a Taylor series expansion of the quantities involved in the equilibrium equation of the system around their mean values. Then the coefficients in the expansions of the left- and right-hand sides are identified and evaluated by perturbation analysis [2]. In the context of finite element analysis for quasi-static linear problems, the equilibrium equation obtained after discretizing the geometry generally reads

$$K \cdot U = F \tag{5.19}$$

Suppose the input parameters used in constructing the stiffness matrix K and the load vector F are varying around their mean. As a consequence, the three quantities appearing in the above equation will also vary around the values K_0, U^0, F_0.

The Taylor series expansions of the terms are

$$K = K_0 + \sum_{i=1}^{N} K_i^I \alpha_i + \frac{1}{2} \sum_{i=1}^{N} \sum_{j=1}^{N} K_{ij}^{II} \alpha_i \alpha_j + o\left(\|\alpha\|^2\right) \tag{5.20}$$

$$U = U^0 + \sum_{i=1}^{N} U_i^I \alpha_i + \frac{1}{2} \sum_{i=1}^{N} \sum_{j=1}^{N} U_{ij}^{II} \alpha_i \alpha_j + o\left(\|\alpha\|^2\right) \tag{5.21}$$

$$F = F_0 + \sum_{i=1}^{N} F_i^I \alpha_i + \frac{1}{2} \sum_{i=1}^{N} \sum_{j=1}^{N} F_{ij}^{II} \alpha_i \alpha_j + o\left(\|\alpha\|^2\right) \tag{5.22}$$

where the first-order coefficients are obtained from the first- and second-order derivatives of the corresponding quantities evaluated at $\alpha = 0$,

$$K_i^I = \frac{\partial K}{\partial \alpha_i}\Big|_{\alpha=0}$$

$$K_{ij}^{II} = \frac{\partial^2 K}{\partial \alpha_i \partial \alpha_j}\Big|_{\alpha=0} \tag{5.23}$$

By identifying similar order coefficients on both sides of the equation, one obtains successively,

$$U^0 = K_0^{-1} \cdot F_0$$

$$U_i^I = K_0^{-1} \cdot (F_i^I - K_i^I \cdot U^0) \tag{5.24}$$

$$U_{ij}^{II} = K_0^{-1} \cdot (F_{ij}^{II} \quad K_i^I \cdot U_j^I \quad K_j^I \cdot U_i^I \quad K_{ij}^{II} \cdot U^0)$$

From these expressions, the statistics of U is readily available from that of α. The second-order estimate of the mean is obtained from

$$E[U] \approx U^0 + \frac{1}{2} \sum_{i=1}^{N} \sum_{j=1}^{N} U_{ij}^{II} Cov[\alpha_i, \alpha_j] \tag{5.25}$$

where the first term U^0 is the first-order approximation of the mean. The first-order estimate of the covariance matrix reads,

$$Cov[U,U] \approx \sum_{i=1}^{N} \sum_{j=1}^{N} U_i^I (U_j^I)^T Cov[\alpha_i, \alpha_j] = \sum_{i=1}^{N} \sum_{j=1}^{N} \frac{\partial U}{\partial \alpha_i}\Big|_{\alpha=0} \frac{\partial U^T}{\partial \alpha_j}\Big|_{\alpha=0} Cov[\alpha_i, \alpha_j] \tag{5.26}$$

Introducing the correlation coefficients of the random variables (α_i, α_j),

$$\rho_{ij} = \frac{Cov[\alpha_i, \alpha_j]}{\sigma_{\alpha_i} \sigma_{\alpha_j}} \tag{5.27}$$

The above equation can be rewritten as

$$Cov[U,U] \approx \sum_{i=1}^{N} \sum_{j=1}^{N} \frac{\partial U}{\partial \alpha_i}\Big|_{\alpha=0} \frac{\partial U^T}{\partial \alpha_j}\Big|_{\alpha=0} \rho_{ij} \sigma_{\alpha_i} \sigma_{\alpha_j} \tag{5.28}$$

It is seen that each term of the summation involves the sensitivity of the response to the parameters a_i as well as the variability of these parameters.

The first-order second-moment perturbation method provides a rapid means of estimating statistical moments of the solution of the stochastic system. If the response random field is approximated as a second-order polynomial in a set of independent Gaussian random variables $\xi_1, \xi_2, \dots \xi_{N_{KL}}$

$$u(x,w) = z_0(x) + \sum_{m=1}^{N_{KL}} \xi_m z_m(x) + \frac{1}{2} \sum_{n=1}^{N_{KL}} \sum_{m=1}^{N_{KL}} \xi_m \xi_n z_{mn}(x) \tag{5.29}$$

With $\langle \xi_m \rangle = 0$ and $\langle \xi_m \xi_n \rangle = \delta_{mn}$, the mean and autocovariance of the response are given by,

$$\bar{u}(x) = z_0(x) + \frac{1}{2}\sum_{n=1}^{N_{KL}}\sum_{m=1}^{N_{KL}}\left\langle \xi_m \xi_n \right\rangle z_{mn}(x) = z_0(x) + \frac{1}{2}\sum_{n=1}^{N_{KL}} z_{mn}(x) \tag{5.30}$$

$$C_{uu}(x,x') = \sum_{n=1}^{N_{KL}}\sum_{m=1}^{N_{KL}}\left\langle \xi_m \xi_n \right\rangle z_m(x) \otimes z_n(x') = \sum_{m=1}^{N_{KL}} z_m(x) \otimes z_m(x') \tag{5.31}$$

Substituting the nodal values corresponding to the expansion,

$$\{u\} = \{z_0\} + \sum_{m=1}^{N_{KL}} \xi_m \{z_m\} + \frac{1}{2}\sum_{n=1}^{N_{KL}}\sum_{m=1}^{N_{KL}} \xi_m \xi_n \{z_{mn}\} \tag{5.32}$$

Ignoring all terms higher than the second order in the random variables and isolating the coefficients of the linearly independent random terms gives

$$[K_0]\{z_0\} = \{f_0\}$$

$$[K_0]\{z_m\} = \{f_m\} - [K_m]\{z_0\}, \quad m = 1,2,\ldots,N_{KL} \tag{5.33}$$

$$[K_0]\sum_{m=1}^{N_{KL}}\{z_{mm}\} = -2\sum_{m=1}^{N_{KL}}[K_m]\{z_m\}$$

The set of $N_{KL}+2$ nodal vectors $\{z_0\}, \{z_m\}(m = 1,2,\cdots,N_{KL})$ and $\sum_{m=1}^{N_{KL}}\{z_{mm}\}$ required to evaluate the mean and autocovariance of the response can be recovered in turn using the Choleski decomposition of the mean stiffness matrix $[K_0]$. The response function $z_0(x)$, $z_m(x)$ and $\sum_{m=1}^{N_{KL}} z_{mm}(x)$ and the corresponding strains $e_0(x)$, $e_m(x)$, $\sum_{m=1}^{N_{KL}} e_{mm}(x)$ may be determined for any required point x from the nodal solution vectors and the finite element shape functions in the usual way. The mean and autocovariance of the strain and stress are then evaluated using the following:

$$\bar{\varepsilon}(x) = e_0(x) + \frac{1}{2}\sum_{m=1}^{N_{KL}} e_{mm}(x)$$

$$C_{\varepsilon\varepsilon}(x,x') = \sum_{m=1}^{N_{KL}} e_m(x) \otimes e_m(x')$$

$$\tag{5.34}$$

$$\bar{\sigma}(x) = D_0(x) \cdot \bar{\varepsilon}(x) + \sum_{m=1}^{N_{KL}} D_m(x) \cdot e_m(x)$$

$$C_{\sigma\sigma}(x,x') = \sum_{m=1}^{N_{KL}}(D_0(x) \cdot e_m(x) + D_m(x) \cdot e_0(x)) \otimes (D_0(x) \cdot e_m(x') + D_m(x) \cdot e_0(x'))$$

5.3   SPECTRAL SFEM

The spectral SFEM aims at discretizing the "random dimension" in a more efficient way using series expansions. Suppose that Young's modulus of the material is
a Gaussian random field [3]. The elasticity matrix in point x can thus be written as

$$D(x, \theta) \equiv H(x, \theta)D_0 \tag{5.35}$$

where D_0 is a constant matrix. The KL expansion of $H(\,)$ is

$$H(x,0) = \mu(x) + \sum_{i=1}^{\infty} \sqrt{\lambda_i} \xi_i(0)\varphi_i(x) \tag{5.36}$$

$$k^e(\theta) = k_0^e + \sum_{i=1}^{\infty} k_i^e \xi_i(\theta) \tag{5.37}$$

where k_0^e is the mean element stiffness matrix and k_i^e are deterministic matrices
obtained by

$$k_i^e = \sqrt{\lambda_i} \int_{\Omega_e} \varphi_i(x)B^T \cdot D_0 \cdot B d\Omega_e \tag{5.38}$$

Assembling the above element contributions eventually gives the stochastic counter-
part of the equilibrium equation,

$$\left[K_0 + \sum_{i=1}^{\infty} K_i \xi_i(\theta) \right] \cdot U(\theta) = F \tag{5.39}$$

In the above equation, K_i are the deterministic matrices obtained by assembling k_i^e in
a way similar to the deterministic case.

By representing the response using the Neumann series, the vector of nodal displacements $U(\theta)$ is formally obtained by inverting. However, no closed-form solution
for such an inverse exists.

$$K_0 \cdot \left[I + \sum_{i=1}^{\infty} K_0^{-1} \cdot K_i \xi_i(\theta) \right] \cdot U(\theta) = F \tag{5.40}$$

which leads to

$$U(\theta) = \left[I + \sum_{i=1}^{\infty} K_0^{-1} \cdot K_i \xi_i(\theta) \right]^{-1} \cdot U^0, \quad U^0 = K_0^{-1} \cdot F \tag{5.41}$$

The Neumann series expansion of the above equation has the form

$$U(\theta) = \sum_{k=0}^{\infty} (-1)^k \left[\sum_{i=1}^{\infty} K_0^{-1} \cdot K_i \xi_i(\theta) \right]^k \cdot U^0 \tag{5.42}$$

whose first terms explicitly write

$$U(\theta) = \left[I - \sum_{i=1}^{\infty} K_0^{-1} \cdot K_i \xi_i(\theta) + \sum_{i=1}^{\infty}\sum_{j=1}^{\infty} K_0^{-1} \cdot K_i \cdot K_0^{-1} \cdot K_j \xi_i(\theta)\xi_j(\theta) + \cdots \right] \cdot U^0 \tag{5.43}$$

Truncating both the KL and Neumann expansions yields an approximate solution for $U(\theta)$. A finite element model of a structure involves a representation of the solid geometry (element mesh), applied loading (boundary conditions), and material properties. When the material properties have random variability, the matrix of constitutive coefficients D, relating stress and strain, $\sigma = D \cdot \varepsilon$, can be defined as a continuous random field $D(x,\omega)$ consistent with a given mean $\bar{D}(x) = \langle D(x,\omega) \rangle$ and autocovariance function,

$$C_{DD}(x,x') = \langle (D(x,\omega) - \bar{D}(x)) \otimes (D(x',\omega) - \bar{D}(x')) \rangle \tag{5.44}$$

where $\otimes$ denotes an outer tensor product. The autocovariance function quantifies the stochastic variability of a random field about its mean and how that variability correlates at different spatial locations. In general, each component of the constitutive material matrix would have its own mean and autocovariance functions, which could be estimated from statistical measurements on samples of the material. However, to simplify the description and application of the method, it is assumed that the constitutive coefficient matrix is of the form $D(x,\omega) = \bar{D}(x) + f(x,\omega)\hat{D}$, where $f(x,\omega)$ is a zero-mean scalar random field and $\hat{D}$ is a constant matrix. Then the autocovariance function is of the form

$$C_{DD}(x,x') = \hat{D} \otimes \hat{D} R_{DD}(x,x') \tag{5.45}$$

where $R_{DD}(x,x')$ is a scalar autocovariance function. The random field $D(x,\omega)$ associated with a given autocovariance function may be represented as a KL expansion (KLE),

$$D(x,\omega) = \bar{D}(x) + \hat{D} \sum_{m=1}^{\infty} \xi_m(\omega) \sqrt{\lambda_m} \Lambda_m(x) \tag{5.46}$$

Defined in terms of a set of eigenvalues λ_m and eigenfunctions $\Lambda_m(x)$ obtained from the spectral decomposition of the scalar autocovariance function,

$$R_{DD}(x,x') = \sum_{m=1}^{\infty} \lambda_m \Lambda_m(x) \Lambda_m(x') \tag{5.47}$$

By solution of the integral eigen-problem,

$$\int_V R_{DD}(x,x')\Lambda_m(x)dV(x) = \lambda_m\Lambda_m(x')$$

$$\int_V \Lambda_m(x)\Lambda_n(x)dV(x) = \delta_{mn}$$

(5.48)

In the KLE, the stochastic variability of $D(x,\omega)$ is expressed in terms of a matrix $\hat{D}$ that incorporates the degree of variability (i.e. the standard deviation) of each component of $D(x,\omega)$, and a set of uncorrelated stochastic functions $\xi_m(\omega)$, $m = 1,2,...$, which are regarded as independent normalized Gaussian random variables with $\langle\xi_m\rangle = 0$ and $\langle\xi_m\xi_n\rangle = \delta_{mn}$. The sequence of λ_m and $\Lambda_m(x)$, $m = 1,2,...$, is ordered in decreasing magnitude of the eigenvalues, so that the infinite series can be truncated after N_{KL} terms to provide a manageable approximation to the random field $D(x,\omega)$.

When a truncated KLE for material properties $D(x,\omega)$ is used in a finite element scheme, there results a system of equations of the form,

$$\sum_{m=0}^{N_{KL}}\xi_m[K_m]\{u\} = \{f\} - \sum_{m=0}^{N_{KL}}\xi_m[\tilde{K}_m]\{\tilde{u}\} = \sum_{m=0}^{N_{KL}}\xi_m\{f_m\}$$

(5.49)

where an extra parameter $\xi_0 = 1$ is defined for notational convenience. The unknowns are contained in the nodal response vector $\{u\}$ and the boundary conditions in the nodal force vectors $\{f_m\}$, $m = 0,1,2,...,N_{KL}$, which are composed of an applied traction/force vector $\{f\}$ and specified nodal response values $\{\tilde{u}\}$. The matrix terms are schematically of the form

$$[K_0] = \int_V \{g'(x)\}\cdot\bar{D}(x)\cdot\{g'(x)\}^T dV(x)$$

$$[K_m] = \int_V \{g'(x)\}\cdot\hat{D}\cdot\{g'(x)\}^T \sqrt{\lambda_m}\Lambda_m(x)dV(x), \quad m = 1,2,...,N_{KL}$$

(5.50)

where $\{g'(x)\}$ involves the shape function derivative associated with strain at x expressed in terms of the response at the nodal locations. The matrix $[K_0]$ is the finite element global stiffness matrix obtained from the mean material properties $\bar{D}(x)$, while the subsequent matrices $[K_m]$, $m = 1,2,...,N_{KL}$ are calculated analogously using the KLE terms $\hat{D}\sqrt{\lambda_m}\Lambda_m(x)$ in place of $\bar{D}(x)$.

Using computer-generated pseudo-random numbers for the Gaussian variables $\xi_1,\xi_2,...,\xi_{N_{KL}}$, the finite element equation system can be used as a basis for Monte Carlo simulation, from which realizations of the stochastic response $u(x,\omega)$ may be computed and used to accumulate statistical moments. The spectral stochastic finite element method provides an alternative solution methodology, in which the random field $u(x,\omega)$ is approximated as a polynomial chaos expansion (PCE) up to the polynomial order p of the form

$$u(x,\omega) = \sum_{j=0}^{N_p} \psi_j(\xi) v_j(x) \tag{5.51}$$

where $\psi_j(\xi)$, $j = 0,1,2,\ldots,N_p$ are chaos polynomials in the N_{KL} random variables $\xi = \{\xi_1,\xi_2,\ldots,\xi_{N_{KL}}\}$, $v_j(x)$ represents a set of spatial functions to be determined, and the number of terms in the PCE of order p in N_{KL} random variables is

$$N_p + 1 = \frac{(p+N_{KL})!}{(p)!(N_{KL})!} \tag{5.52}$$

Chaos polynomials are defined to be orthogonal with respect to expectation,

$$\langle \psi_j(\xi)\psi_k(\xi) \rangle = 0 \quad \text{if } j \neq k \tag{5.53}$$

For convenience, the first $N_{KL} + 1$ polynomials are of the order 0 term $\psi_0(\xi) = 1$ and of the order 1 term $\psi_j(\xi) = \xi_j$ ($j = 1,2,\ldots,N_{KL}$). For Gaussian random variables ξ, the sequence of chaos polynomials may be constructed as products of one-dimensional monic Hermite polynomials. In a finite element scheme, the nodal response vector corresponding to the expansion is

$$\{u(\omega)\} = \sum_{j=0}^{N_p} \psi_j(\xi)\{v_j\} \tag{5.54}$$

where each $v_j(x) = \{g(x)\}^T \{v_j\}$ is represented in terms of nodal values $\{v_j\}$ and finite element shape functions $\{g(x)\}$. Then,

$$\sum_{j=0}^{N_p}\sum_{m=0}^{N_{KL}} \xi_m \psi_j(\xi)[K_m]\{v_j\} = \sum_{m=0}^{N_{KL}} \xi_m \{f_m\} \tag{5.55}$$

The orthogonality of the chaos polynomials may be exploited in a Galerkin procedure to reduce to a block-sparse system of deterministic equations in the $(N_p + 1)$ sets of nodal values $\{v_j\}$, $j = 0,1,2,\ldots N_p$. Multiplying by $\psi_k(\xi)$ and taking expectation gives

$$\sum_{j=0}^{N_p}\sum_{m=0}^{N_{KL}} c_{mjk}[K_m]\{v_j\} = \sum_{m=0}^{N_{KL}} \delta_{km}\{f_m\} \quad k = 0,1,2,\ldots,N_p \tag{5.56}$$

where the coefficients

$$c_{mjk} = \langle \xi_m \psi_j(\xi)\psi_k(\xi) \rangle \tag{5.57}$$

Once the solution vectors $\{v_j\}$ have been determined, the mean and autocovariance of the response may be obtained as

$$\bar{u}(x) = \sum_{j=0}^{N_p} \langle \psi_j(\xi) \rangle v_j(x) = v_0(x)$$

(5.58)

$$C_{uu}(x,x') = \sum_{j=1}^{N_p} \langle (\psi_j(\xi))^2 \rangle v_j(x) \otimes v_j(x')$$

where $v_j(x) = \{g(x)\}^T \{v_j\}$ is the contribution to the response at location x from the jth nodal solution vector and $\langle (\psi_j(\xi))^2 \rangle$ can be evaluated.

The strain and stress at a location x within the structure are given by

$$\varepsilon(x,\omega) = \sum_{j=0}^{N_p} \psi_j(\xi) \varepsilon_j(x)$$

(5.59)

$$\sigma(x,\omega) = D(x,\omega) \cdot \varepsilon(x,\omega) = \sum_{m=0}^{N_{KL}} \sum_{j=0}^{N_p} \xi_m \psi_j(\xi) D_m(x) \cdot \varepsilon_j(x)$$

where $\varepsilon_j(x) = \{g'(x)\}^T \{v_j\}$ are strain-type values at x derived from the jth nodal solution vector and the effective material coefficients are given by $D_0(x) = \bar{D}(x)$ and $D_m(x) = \hat{D}\sqrt{\lambda_m}\Lambda_m(x)$ for $m = 1,2,\ldots,N_{KL}$. Defining $\sigma_{mj}(x) = D_m(x) \cdot \varepsilon_j(x)$, the mean and autocovariance of strain and stress can be obtained using

$$\bar{\varepsilon}(x) = \sum_{j=0}^{N_p} \langle \psi_j(\xi) \rangle \varepsilon_j(x) = \varepsilon_0(x)$$

$$C_{\varepsilon\varepsilon}(x,x') = \sum_{j=1}^{N_p} \langle (\psi_j(\xi))^2 \rangle \varepsilon_j(x) \otimes \varepsilon_j(x')$$

(5.60)

$$\bar{\sigma}(x) = \sum_{m=0}^{N_{KL}} \sum_{j=0}^{N_p} \langle \xi_m \psi_j(\xi) \rangle \sigma_{mj}(x) = \sum_{m=0}^{N_{KL}} \sigma_{mm}(x)$$

$$C_{\sigma\sigma}(x,x') = \sum_{k=0}^{N_p} \sum_{j=0}^{N_p} \sum_{n=0}^{N_{KL}} \sum_{m=0}^{N_{KL}} (X_{mnjk}\sigma_{mj}(x) \otimes \sigma_{nk}(x')) - \bar{\sigma}(x) \otimes \bar{\sigma}(x')$$

The nonzero coefficients $X_{mnjk} = \langle \xi_m \xi_n \psi_j(\xi) \psi_k(\xi) \rangle$ can be identified and calculated analogously to the coefficients c_{mjk} using a recurrence formula for one-dimensional Hermite polynomials [4].

5.4 NEUMANN SFEM

Consider a linear elastic static system with Young's modulus modeled as a random field. By using the KL expansion or the Fourier–Karhunen–Loève expansion, random Young's modulus can be expressed as

$$E(x,\omega) = E_0(x) + \sum_i \sqrt{\lambda_i} \alpha_i(w) \psi_i(x) \tag{5.61}$$

where $E(x,\omega)$ denotes random Young's modulus at point x, $E_0(x)$ is the expectation of $E(x,\omega)$, $\alpha_i(w)$ represents uncorrelated random variables, and λ_i and $\psi_i(x)$ are deterministic eigenvalues and eigenfunctions corresponding to the covariance function of the source field $E(x,\omega)$. If $E(x,\omega)$ is modeled as a Gaussian field, $\alpha_i(w)$ becomes independent standard Gaussian random variables.

Following the standard finite element discretization procedure, a stochastic system of linear algebraic equations can be built,

$$\left(A_0 + \sum_i \alpha_i A_i \right) u = f \tag{5.62}$$

In the above equation, A_0 is the conventional FE stiffness matrix, which is symmetric, positive definite, and sparse while A_i represents symmetric and sparse matrices. The deterministic matrices A_i share the same entry pattern as the stiffness matrix A_0, but they are usually not positive definite.

In the Neumann expansion method, the solution of the above equation can be written as

$$u = (A_0 + \Delta A)^{-1} f = (I + B)^{-1} A_0^{-1} f \tag{5.63}$$

where $\Delta A = \sum_i \alpha_i A_i$, $B = A_0^{-1} \Delta A$ and I is the identity matrix. According to the Neumann expansion,

$$(I + B)^{-1} = I - B + B^2 - B^3 + \cdots \tag{5.64}$$

Then,

$$u = u_0 - B u_0 + B^2 u_0 - B^3 u_0 + \cdots \tag{5.65}$$

where $u_0 = A_0^{-1} f$. The spectral radius of matrix B must be smaller than 1 to ensure the convergence. The Neumann expansion can be generalized to the inverse of multiple matrices,

$$\left(I + \sum_i \alpha_i B_i \right)^{-1} = I - \sum_i \alpha_i B_i + \sum_i \sum_j \alpha_i \alpha_j B_i B_j - \tag{5.66}$$

The generalized Neumann expansion can be proved as follows:

$$\left(I + \sum_i \alpha_i B_i\right)\left(I + \sum_i \alpha_i B_i\right)^{-1} = I - \sum_i \alpha_i B_i + \sum_i \sum_j \alpha_i \alpha_j B_i B_j - \ldots + \alpha_i B_i$$

$$- \sum_i \alpha_i \alpha_j B_i B_j + \alpha_2 B_2 - \sum_i \alpha_2 \alpha_i B_2 B_i + \ldots + \alpha_n B_n - \sum_i \alpha_n \alpha_i B_n B_i + \ldots = I \quad (5.67)$$

where n is the total number of B_i.

The Neumann expansion and the generalized Neumann expansion are equivalent in mathematics and only differ in organizing the matrix operations. This difference can lead to a significant acceleration in computing the solution. Specifically, let $B_i = A_0^{-1} A_i$,

$$u = \left(A_0 + \sum_i \alpha_i A_i\right)^{-1} f = \left(I + \sum_i \alpha_i A_0^{-1} A_i\right)^{-1} A_0^{-1} f$$

$$= u_0 - \sum_i \alpha_i A_0^{-1} A_i u_0 + \sum_i \sum_j \alpha_j \alpha_i A_0^{-1} A_j A_0^{-1} A_i u_0 - \ldots$$

$$= u_0 - \sum_i \alpha_i u_i + \sum_i \sum_j \alpha_i \alpha_j u_{ij} - \ldots \quad (5.68)$$

in which

$$u_i = A_0^{-1} A_i u_0$$
$$u_{ij} = A_0^{-1} A_i A_0^{-1} A_j u_0 \quad (5.69)$$

The convergence condition for the Neumann expansion is

$$\rho\left(\sum_i \alpha_i A_0^{-1} A_i\right) < 1 \quad (5.70)$$

It should be noted that

$$\rho\left(\sum_i \alpha_i A_0^{-1} A_i\right) = \rho\left(\sum_i \alpha_i C^{-1} A_i C^{-1}\right) = \left\|\sum_i \alpha_i C^{-1} A_i C^{-1}\right\|_{m2} \quad (5.71)$$

where $\|\cdot\|_{m2}$ denotes the spectral norm of a matrix, and the symmetric and positive definite matrix C is defined as

$$A_0 = C^2 \quad (5.72)$$

5.5   FINITE ELEMENT RELIABILITY ANALYSIS

Reliability methods aim at evaluating the probability of failure of a system whose modeling takes into account randomness. Classically, the system is decomposed into components and the system failure is defined by various scenarios about the joint failure of components. Thus the determination of the probability of failure of each component is of paramount importance.

Early work in structural reliability aimed at determining the failure probability in terms of the second-moment statistics of the resistance and demand variables. Suppose these are lumped into two random variables denoted by R and S, respectively. The safety margin is defined by

$$Z = R - S \tag{5.73}$$

Cornell's reliability index is then defined by

$$\beta_C = \frac{\mu_z}{\sigma_z} \tag{5.74}$$

It can be given the following interpretation, if R and S were to be jointly normal, so would be Z. The probability of failure of the system would then be

$$P_f = P(Z \le 0) = P\left(\frac{Z - \mu_Z}{\sigma_Z} \le -\frac{\mu_Z}{\sigma_Z}\right) \equiv \Phi(-\beta_C) \tag{5.75}$$

where $\Phi(\cdot)$ is the standard normal cumulative distribution function. In this case, β_C can be described as a function of the second-moment statistics of R and S,

$$\beta_C = \frac{\mu_R - \mu_S}{\sqrt{\sigma_R^2 + \sigma_S^2 - 2\rho_{RS}\sigma_R\sigma_S}} \tag{5.76}$$

Consider a general case where Z is actually a limit state function,

$$Z = g(S) \tag{5.77}$$

The mean μ_S and covariance matrix $\displaystyle\sum_{SS}$ are known. The mean and covariance of Z are not available in the general case where $g(S)$ is nonlinear. Using the Taylor expansion around the mean value of S,

$$Z = g(\mu_s) + (\nabla_s g)_{S=\mu_s}^T \cdot (S - \mu_s) + o\left(\left\|(S - \mu_s)^2\right\|\right) \tag{5.78}$$

The following first-order approximations are obtained,

$$\mu_z \approx g(\mu_s)$$

$$\sigma_z^2 \approx (\nabla_s g)_{S=\mu_s}^T \cdot \sum_{SS} \cdot (\nabla_s g)_{S=\mu_s} \tag{5.79}$$

This procedure leads to the definition of the so-called mean value first-order second-moment reliability index,

$$\beta_{\text{MVFOSM}} = \frac{g(\mu_s)}{\sqrt{(\nabla_s g)^T_{S=\mu_s} \cdot \sum_{ss} \cdot (\nabla_s g)_{S=\mu_s}}} \tag{5.80}$$

The main problem with this reliability index is that it is not invariant with respect to changing the limit state function for an equivalent one. Variations can be important in some problems.

Consider a probabilistic transformation of the basic random variable X,

$$Y = y(X) \tag{5.81}$$

such that Y is a Gaussian random vector with zero mean and unit covariance matrix.

The exact expression of the probability transformation $Y = y(X)$ depends on the joint PDF of X. Several cases sorted by an ascending order of difficulty are listed in the sequel as examples, X is a Gaussian random vector with mean μ_χ and covariance matrix $\sum_{\chi\chi}$. The diagonalization of the symmetric positive definite matrix $\sum_{\chi\chi}$ allows us to write

$$\chi = A \cdot Y + \mu_\chi \tag{5.82}$$

where A is obtained by the Cholesky decomposition of $\sum_{\chi\chi}$,

$$\sum_\chi = A \cdot A^T \tag{5.83}$$

The probability transformation and its Jacobian then write,

$$y(\chi) = A^{-1}(\chi - \mu_\chi)$$
$$J_{y,\chi} = A^{-1} \tag{5.84}$$

χ is a vector of independent nonnormal variables whose PDF $f_i(x_i)$ and CDF $F_i(x_i)$ are given. The probability transformation, in this case, is diagonal,

$$y_i = \Phi^{-1}[F_i(x_i)], \quad i = 1,...N \tag{5.85}$$

Its Jacobian reads

$$J_{y,\chi} = diag\left(\frac{f(x_i)}{\varphi(y_i)}\right) \tag{5.86}$$

χ is a vector of dependent nonnormal variables. In many applications, the joint PDF of these random variables is not known. The available information is often limited to the marginal distributions and correlation matrix R, whose coefficients read,

$$\rho_{ij} = \frac{Cov[\chi_i, \chi_j]}{\sigma_i \sigma_j} \tag{5.87}$$

The Morgenstern model: it is limited to small correlation ($|\rho_{ij}| < 0.3$) and the closed-form expression for the joint PDF and CDF become tedious to manage when dealing with a large number of random variables.

The Nataf model: it is defined in a convenient closed form for any number of random variables and complies with almost any valid correlation structure.

Due to these characteristics, only the Nataf model is presented now. From the marginal PDF of X_i, the following random vector Z is defined,

$$Z_i = \Phi^{-1}[F_i(\chi_i)] \tag{5.88}$$

Assuming that Z is a Gaussian standard normal vector with yet to be computed correlation matrix R_0, its joint PDF is given by

$$f_Z(z) = \varphi_n(z, R_0) \equiv \frac{1}{(2\pi)^{n/2}\sqrt{\det R_0}} \exp\left(-\frac{1}{2}z^T \cdot R_0^{-1} \cdot z\right) \tag{5.89}$$

Using the inverse transformation, the joint PDF of χ then reduces to

$$f_\chi(\chi) = f_1(x_1)\cdots f_n(x_n)\frac{\varphi_n(z, R_0)}{\varphi(z_1)\cdots\varphi(z_n)} \tag{5.90}$$

To complete the definition, the correlation matrix R_0 should finally be chosen such that the correlation coefficient of any pair matches the prescribed correlation coefficient ρ_{ij}. This condition leads to the following implicit equation in $\rho_{o,ij}$:

$$\rho_{ij} = \int_{-\infty}^{\infty}\int_{-\infty}^{\infty}\left(\frac{x_i - \mu_i}{\sigma_i}\right)\left(\frac{x_j - \mu_j}{\sigma_j}\right)\varphi_2\left(z_i, z_j, \rho_{o,ij}\right)dz_i dz_j \tag{5.91}$$

From a reliability point of view, assuming the basic variables χ have a Nataf joint PDF, vector Z defined is Gaussian with zero mean and correlation matrix $R_0 = L_0 \cdot L_0^T$. The probability transformation to the standard normal space thus is

$$y(X) = L_0^{-1} \cdot Z = L_0^{-1} \cdot \left\{\Phi^{-1}[F_1(x_1)],\ldots,\Phi^{-1}[F_n(x_n)]\right\}^T \tag{5.92}$$

Its Jacobian is

$$J_{y,\chi} = L_0^{-1} \cdot diag\left(\frac{f(x_i)}{\varphi(y_i)}\right) \tag{5.93}$$

The Rosenblatt transformation is an alternative possibility when conditional PDFs of χ are known. It is defined as follows:

$$\begin{cases} y_1 = \Phi^{-1}\left[F(x_1)\right] \\ y_2 = \Phi^{-1}\left[F(x_2|x_1)\right] \\ \vdots \\ y_n = \Phi^{-1}\left[F(x_n|x_1,\ldots,x_n)\right] \end{cases} \tag{5.94}$$

Unfortunately, it is not invariant by the permutation of the variables χ_i.

The mapping of the limit state function onto the standard normal space by using the probabilistic transformation is described by

$$g(S) = G(Y) \tag{5.95}$$

Hence, the probability of failure can be rewritten as

$$P_f = \int_{G(y)\leq 0} \varphi(y)\,dy \tag{5.96}$$

where $\varphi(Y)$ denotes the standard normal PDF of Y,

$$\varphi = \frac{1}{(2\pi)^{n/2}} \exp\left(-\frac{1}{2}\|y\|^2\right) \tag{5.97}$$

This PDF has two interesting properties, namely, it is rotationally symmetric and decays exponentially with the square of the norm $\|y\|$. Thus, the points making significant contributions to the integral are those with the nearest distance to the origin of the standard normal space. This leads to the definition of the reliability index β,

$$\beta = \alpha^T \cdot y^*$$
$$y^* = \arg\min\left\{\|y\| \;\; |G(y) \leq 0\right\} \tag{5.98}$$

This quantity is obviously invariant under changes in the parametrization of the limit state function, since it has an intrinsic definition, the distance of the origin to the limit state surface.

REFERENCES

1. Stefanou, G. The stochastic finite element method: Past, present and future. *Computer Methods in Applied Mechanics and Engineering*, **2009**, 198(9–12), 1031–1051.
2. Xia, B., Yu, D., Liu, J. Transformed perturbation stochastic finite element method for static response analysis of stochastic structures. *Finite Elements in Analysis and Design*, **2014**, 79, 9–21.

3. Zakian, P., Khaji, N. A stochastic spectral finite element method for wave propagation analyses with medium uncertainties. *Applied Mathematical Modelling*, **2018**, 63, 84–108.

4. Ngah, M.F, Young, A. Application of the spectral stochastic finite element method for performance prediction of composite structures. *Composite Structures*, **2007**, 78(3), 447–456.

6 Machine Learning Methods

6.1 ARTIFICIAL NEURAL NETWORKS

Artificial neural networks (ANNs) are computing systems that imitate the biological neural networks that constitute animal brains. They are the foundation of artificial intelligence and solve problems that would prove impossible or difficult by human or statistical standards. ANNs learn by processing original examples, each of which contains a known "input" and " related result", forming the probability-weighted associations between the two [1].

An ANN consists of a collection of simulated neurons. Each neuron is a node which is connected to other nodes via links that correspond to biological axon–synapse–dendrite connections. Each link has a weight, which determines the strength of the influence of one node on another (Figure 6.1).

The training of an ANN that forms a given example is usually conducted by determining the difference between the processed outputs of the network (often a prediction) and the target results. To minimize the relative error, the weighted associations are adjusted according to the learning rules. Successive adjustments will cause the ANNs to produce output which is increasing approximated to the target output.

6.1.1 Components of ANN

1. **Neurons**

    ANNs are composed of artificial neurons which are conceptually derived from biological neurons. Each artificial neuron has inputs and produces a single output which can be sent to multiple other neurons. The inputs can be feature values of a sample of external data, or they can be the outputs of other neurons. The outputs of the final output neurons of the neural net accomplish the approximation.

2. **Connections and weights**

    The network consists of connections, with each connection providing the output of one neuron as an input to another neuron. Each connection is assigned a weight that represents its relative importance. A given neuron can have multiple input and output connections.

3. **Propagation function**

    The propagation function computes the input to a neuron from the outputs of its predecessor neurons and their connections as a weighted sum. A bias term can be added to the result of the propagation.

DOI: 10.1201/9781003226628-7

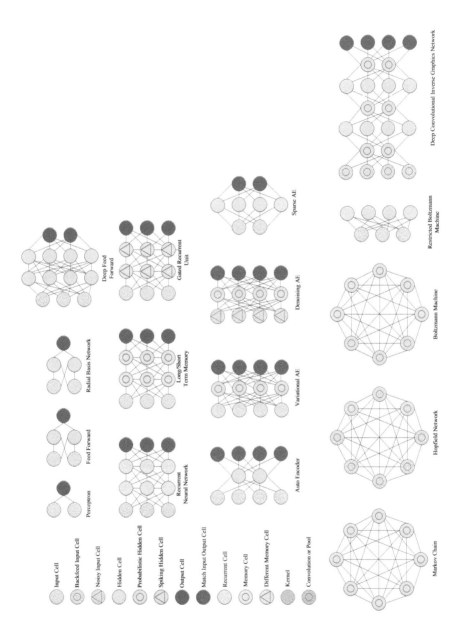

FIGURE 6.1 The main ANN methods.

4. **Organization**

The neurons are typically organized into multiple layers, especially in deep learning. Neurons of one layer connect only to neurons of the immediately preceding and immediately following layers. The layer that receives external data is the input layer. The layer that produces the ultimate result is the output layer. In between them are zeroes or more hidden layers. Single-layer and un-layered networks are also used. Between two layers, multiple connection patterns are possible. They can be fully connected, with every neuron in one layer connecting to every neuron in the next layer. They can be pooling, where a group of neurons in one layer connect to a single neuron in the next layer, thereby reducing the number of neurons in that layer. Neurons with only such connections form a directed acyclic graph and are known as feedforward networks. Alternatively, networks that allow connections between neurons in the same or previous layers are known as recurrent networks.

5. **Hyper-parameter**

A hyper-parameter is a constant parameter whose value is set before the learning process begins. The values of parameters are derived via learning. Examples of hyper-parameters include learning rate, the number of hidden layers, and batch size. The values of some hyper-parameters can be dependent on those of other hyper-parameters. For example, the size of some layers can depend on the overall number of layers.

6. **Learning**

Learning is the adaptation of the network to better handle a task by considering sample observations. Learning involves adjusting the weights and optional thresholds of the network to improve the accuracy of the result. This is done by minimizing the observed errors. Learning is complete when examining additional observations does not usefully reduce the error rate. Even after learning, the error rate typically does not reach 0. If after learning, the error rate is too high, the network typically must be redesigned. Practically this is done by defining a cost function that is evaluated periodically during learning. As long as its output continues to decline, learning continues. The cost is frequently defined as a statistic whose value can only be approximated. The outputs are actually numbers, so when the error is low, the difference between the output and the correct answer is small. Learning through attempts reduces the total of the differences across the observations. Most learning models can be viewed as a straightforward application of the optimization theory and statistical estimation.

7. **Learning rate**

The learning rate defines the size of the corrective steps that the model takes to adjust for errors in each observation. A high learning rate shortens the training time, but with a lower ultimate accuracy, while a lower learning rate task takes longer, but with the potential for greater accuracy. Optimization is primarily aimed at speeding up error minimization, while other improvements mainly try to increase reliability. To avoid oscillation inside the network such as alternating connection weights, and to improve the rate of convergence, refinements use an adaptive learning rate that increases or decreases as appropriate. The concept of momentum allows the balance between the gradient and the previous change to be weighted such that the weight adjustment depends to some degree on the previous change. A momentum close to 0 emphasized the gradient, while a value close to 1 emphasizes the last change.

8. **Cost function**

While it is possible to define a cost function, frequently the choice is determined by the desirable properties of the function, such as convexity, or because it arises from the model.

9. **Backpropagation**

Backpropagation is a method used to adjust the connection weights to compensate for each error found during learning. The error amount is effectively divided among the connections. Technically, backpropagation calculates the gradient (the derivative) of the cost function associated with a given state with respect to the weights. The weight updates can be done via stochastic gradient descent or other methods.

6.1.2   Learning Paradigms

Supervised learning uses a set of paired inputs and desired outputs. The learning task is to produce the desired output for each input. In this case, the cost function is related to eliminating incorrect deductions. A commonly used cost is the mean-squared error (MSE), which tries to minimize the average squared error between the network output and the desired output. Tasks suited for supervised learning are pattern recognition and regression. Supervised learning is also applicable to sequential data.

In **unsupervised learning**, input data are given along with the cost function, some functions of the data, and the network output. The cost function is dependent on the task (the model domain) and any a priori assumptions. Minimizing this cost produces a value that is equal to the mean of the data. The cost function can be much more complicated. Its form depends on the application. Tasks that fall within the paradigm of unsupervised learning are in general estimation problems, and the application includes clustering, the estimation of statistical distribution, compression, and filtering.

In **reinforcement learning**, the aim is to weigh the network to perform actions that minimize long-term costs. At each point in time, the agent performs an action and the environment generates an observation and an instantaneous cost, according to some rules. The rules and the long-term cost usually only can be estimated. At any juncture, the agent decides whether to explore new actions to uncover their costs or to exploit prior learning to proceed more quickly.

6.1.3   Theoretical Properties

6.1.3.1   Computational Power

The multilayer perceptron is a universal function approximator, as proven by the universal approximation theorem. However, the proof is not constructive regarding the number of neurons required, the network topology, the weights, and the learning parameters.

A specific recurrent architecture with rational-valued weights has the power of a universal Turing machine, using a finite number of neurons and standard linear connections. Furthermore, the use of irrational values for weights results in a machine with super Turing power.

6.1.3.2 Capacity

It is related to the amount of information that can be stored in the network and to the notion of complexity. Two notions of capacity are known by the community.

6.1.3.3 Convergence

Models may not consistently converge on a single solution, firstly because local minima may exist, depending on the cost function and the model. Secondly, the optimization method used might not guarantee to converge when it begins far from any local minimum. Thirdly, for sufficiently large data or parameters, some methods become impractical.

The convergence behavior of certain types of ANN architectures is more understood than others. When the width of the network approaches infinity, the ANN is well described by its first-order Taylor expansion throughout training and so inherits the convergence behavior of affine models. Another example is when parameters are small, it is observed that ANNs often fit target functions from low to high frequencies. This phenomenon is the opposite of the behavior of some well-studied iterative numerical schemes such as the Jacobi method.

6.1.3.4 Generalization and Statistics

Applications whose goal is to create a system that generalizes well to unseen examples face the possibility of over-training. This arises in convoluted or over-specified systems when the network capacity significantly exceeds the needed free parameters. Two approaches address over-training. The first is to use cross-validation and similar techniques to check for the presence of over-training and to select hyper-parameters to minimize the generalization error.

The second is to use some form of regularization. This concept emerges in a probabilistic framework, where regularization can be performed by selecting a larger prior probability over simpler models, but also in the statistical learning theory, where the goal is to minimize over two quantities, the empirical risk and structural risk, which roughly corresponds to the error over the training set and the predicted error in unseen data due to overfitting.

Supervised neural networks that use a MSE cost function can use formal statistical methods to determine the confidence of the trained model. The MSE on the validation set can be used as an estimate for variance. This value can then be used to calculate the confidence interval of network output, assuming a normal distribution. A confidence analysis made this way is statistically valid as long as the output probability distribution stays the same and the network is not modified [2].

6.2   RADIAL BASIS NETWORK

In terms of the direction of information flow, the radial basis function (RBF) neural network consists of an input layer, a hidden layer, and an output layer. The activation function in its hidden layer is Gaussian and is assumed as $\alpha_t(X), t = 1, 2, \ldots k$. The output value of the t-th hidden neuron can be expressed as

$$\alpha_t(X) = \exp(-\|X - \mu_t\| / \sigma_t^2) \tag{6.1}$$

where $X = (x_1, x_2, \ldots x_m)^T$ and μ_t denote the input of the network and the center vector of the t-th hidden neuron, respectively, $\|X - \mu_t\|$ is the Euclidean distance between X and μ_t, and σ_t is the radius or width of the t-th hidden neuron.

A single-output RBF neural network is a linear combination of units on the hidden layer, which can be described by

$$y = \sum_{t=1}^{k} w_t \alpha_t(X) \tag{6.2}$$

where $W = [w_1, w_2, \ldots, w_k]$ represents the connecting weights between the hidden layer and the output layer.

The main procedures of using the RBF neural network are as follows:

1. Determining sampled points of input parameters. The selected sampled points should reflect the distribution of modeling parameters and have a reasonable quantity.
2. Calculating analytical responses corresponding to the sampled points.
3. Training an RBF neural network, the training of a neural network is implemented to adjust the weights between the hidden layer and the output layer via the training dataset. The least squares method is employed to determine these weights.
4. Verifying the accuracy of the trained network. The accuracy of the trained network is verified by solving N times with new randomly selected sampled points in the design space and outputs of the trained network are compared with the corresponding results.

6.3   BACKPROPAGATION NEURAL NETWORK

The input value of the backpropagation neural network is expressed in $X = (x_1, x_2, \ldots x_m)^T$ and the predicted value is expressed in $Y = (y_1, y_2, \ldots y_m)$. The training process of the backpropagation neural network is as follows:

1. Initializing the neural network. The values of n, T, and m are determined according to the actual prediction requirements, and the hidden layer threshold a and the output layer threshold b are initialized. Then, the learning rate of the neural network and the excitation function of the neuron are determined. The implicit layer excitation function f selected for example is

$$f(x) = 1/(1 + e^{-x}) \tag{6.3}$$

2. Calculating the output of the hidden layer. On the premise that X, w_{ij}, and a are known, the output H of the hidden layer can be obtained by calculating

$$H_j = f\left(\sum_{i=1}^{n} w_{ij} x_i - a_j\right) \quad j = 1, 2, \ldots, l \tag{6.4}$$

In the formula, L is the number of nodes in the hidden layer.

3. Calculating the output of the output layer. From H, w_{jk}, and b, the value of predicted output O can be obtained.

$$O_k = \sum_{j=1}^{l} H_j w_{jk} - b_k \quad k = 1, 2, \ldots, m \tag{6.5}$$

4. Calculating the prediction error. The model prediction error is obtained by subtracting O from the expected output Y.

$$e_k = Y_k - O_k, \quad k = 1, 2, \ldots, m \tag{6.6}$$

5. Updating weights. Update w_{ij} and w_{jk} according to e. the expressions are as follows:

$$w_{ij} = w_{ij} + \eta H_j (1 - H_j) x(i) \sum_{k=1}^{m} w_{jk} e_k \quad i = 1, 2, \ldots, n; \quad j = 1, 2, \ldots, l \tag{6.7}$$

$$w_{jk} = w_{jk} + \eta H_j e_k \quad j = 1, 2, \ldots l; \quad k = 1, 2, \ldots, m$$

In the formula, η is the learning rate.

6. Updating threshold. Update a and b according to e.

$$a_j = a_j + \eta H_j (1 - H_j) x(i) \sum_{k=1}^{m} w_{jk} e_k \quad j = 1, 2, \ldots, l \tag{6.8}$$

$$b_k = b_k + e_k \quad k = 1, 2, \ldots, m$$

7. Judging whether the model converges or not, if not, go back to step 2. Repeat the above steps until the end of the iteration.

6.4 RESTRICTED BOLTZMANN MACHINE

Restricted Boltzmann machine (RBM) is a probabilistic graphical model which can be explained by a stochastic neural network. An RBM can be used to learn a probability distribution over a set of inputs.

The topology of RBM is a two-layer graph. The underlying structure of the network is used to receive the input data, which is called the visible layer represented by $V = [v_1, v_2, \ldots, v_{n_v}]$, and the upper structure of the network is used to generate a new feature vector, which is called the hidden layer represented by $H = [h_1, h_2, \ldots, h_{n_h}]$. For simplicity, the data of the visible and hidden layers in the RBM are assumed to be subject to Bernoulli distribution. The discrete RBM only takes value 0 or 1, and its training is an unsupervised process during which only the feature vector is needed but the tag data are not required.

Given a training set $S = \left\{ v^1, v^2, \ldots, v^N \right\}$ that contains N samples, where $v^m = \left\{ v_1^m, v_2^m, \ldots, v_{nv}^m \right\}$ $m = 1, \ldots, N$ is the m-th training sample. Then, RBM can be considered as an energy model. Given a set of network states (v, h), the RBM network can be corresponding to an energy value through

$$E_\theta(v, h) = -\sum_{j=1}^{n_v} a_j v_j - \sum_{i=1}^{n_h} b_i h_i - \sum_{i=1}^{n_h} \sum_{j=1}^{n_v} h_i w_{i,j} v_j \tag{6.9}$$

where $a = [a_1, a_2, \ldots, a_{n_v}] \in R^{n_v}$ is the visible layer bias, $b = [b_1, b_2, \ldots, b_{n_h}] \in R^{n_h}$ is the hidden layer bias, and $w = [w_{i,j}] \in R^{n_v \times n_h}$ is the weight matrix.

The joint probability distribution of a set of visible and hidden states can be obtained by

$$p_\theta(v, h) = \frac{1}{Z_\theta} e^{-E_\theta(v, h)} \tag{6.10}$$

where $Z_\theta = \sum_{v,h} e^{-E_\theta(v, h)}$ is the normalization factor. The likelihood function $p(v|\theta)$ can be derived by using the joint probability distribution as follows:

$$p(v|\theta) = \frac{1}{Z_\theta} \sum_h e^{-E(v, h|\theta)} \tag{6.11}$$

The training principle for RBM, a free energy function, is defined as

$$\text{Free Entropy}(v) = -\ln \sum_h e^{-E(v, h)} \tag{6.12}$$

Based on the above energy function, the joint probability density distribution can be rewritten as

$$p(v|\theta) = \frac{1}{Z_\theta} e^{-\text{Free Entropy}(v)} \tag{6.13}$$

Taking logarithm for both sides of equality,

$$\ln p(v|\theta) = -\text{Free Entropy}(v) - \ln Z_\theta \tag{6.14}$$

Taking a summation of the above equation for all vectors,

$$\ln \prod_{v} p(v|\theta) = -\sum_{v} \text{Free Entropy}(v) - \ln \prod_{v} Z_\theta \qquad (6.15)$$

There exists a negative relationship between the free energy and the likelihood function $\ln \prod_{v} p(v|\theta)$ for an RBM. Based on the well-known principle of minimum free energy in physical energy systems, the free energy term is supposed to be minimum. The likelihood function is represented as

$$\ln L_{\theta,S} = \ln \prod_{m=1}^{N} p(v^m) = \sum_{m=1}^{N} \ln p(v^m) \qquad (6.16)$$

The purpose of training RBM is to get the optimal value of the parameter θ, that is,

$$\theta^* = \arg \max(\ln L_{\theta,S}) \qquad (6.17)$$

where θ^* is the optimal value that makes the free energy of the RBM system to be minimum. The gradient descent technique can be used to find the maximum of $\ln L_{\theta,S}$ with respect to the parameter θ. Taking the partial derivative,

$$\frac{\partial \ln L_{\theta,S}}{\partial \theta} = \sum_{m=1}^{N} \frac{\partial \ln p(v^m)}{\partial \theta} \qquad (6.18)$$

Replacing the parameter θ with (W, a, b), the following formulas can be obtained:

$$\begin{cases} \dfrac{\partial \ln L_{\theta,S}}{\partial w_{i,j}} = \sum_{m=1}^{N} \left[p(h_i = 1|v^m)v_j^m - \sum_{v} p(v)p(h_i = 1|v)v_j \right] \\[2mm] \dfrac{\partial \ln L_{\theta,S}}{\partial a_j} = \sum_{m=1}^{N} \left[v_j^m - \sum_{v} p(v)v_j \right] \\[2mm] \dfrac{\partial \ln L_{\theta,S}}{\partial b_i} = \sum_{m=1}^{N} \left[p(h_i = 1|v^m) - \sum_{v} p(v)p(h_i = 1|v) \right] \end{cases} \qquad (6.19)$$

The conditional probability is given by

$$\begin{cases} p(h_i = 1|v) = \text{sigmoid}\left(b_i + \sum_{j=1}^{n_v} w_{i,j}v_j \right) \\[4mm] p(v_j = 1|h) = \text{sigmoid}\left(a_j + \sum_{i=1}^{n_h} w_{i,j}h_i \right) \end{cases} \qquad (6.20)$$

It is still unable to update the parameters based on these gradient formulas, since the complexity of computing $\sum_v$ is very high. An efficient approximation method is the CD algorithm proposed by Hinton. The gradient formulas included in the CD algorithm are listed as follows:

$$\begin{cases} \dfrac{\partial \ln L_{\theta,S}}{\partial w_{i,j}} \approx \sum_{m=1}^{N} \left[p(h_i = 1 | v^{(m,0)}) v_j^{(m,0)} - p(h_i = 1 | v^{(m,k)}) v_j^{(m,k)} \right] \\[2em] \dfrac{\partial \ln L_{\theta,S}}{\partial a_j} \approx \sum_{m=1}^{N} \left[v_j^{(m,0)} - v_j^{(m,k)} \right] \\[2em] \dfrac{\partial \ln L_{\theta,S}}{\partial b_i} \approx \sum_{m=1}^{N} \left[p(h_i = 1 | v^{(m,0)}) - p(h_i = 1 | v^{(m,k)}) \right] \end{cases} \tag{6.21}$$

where k is the number of sampling times in the CD algorithm, which is usually equal to 1, and 0 represents the starting point of sampling.

6.5 HOPFIELD NEURAL NETWORK

The input and the output of complex-valued multistate neurons are complex numbers. The output of a complex-valued neuron is determined by an activation function. Let K be a positive integer greater than 1. In this brief, for a complex number I, we define an activation function $f(I)$, In the case of $\arg(I) = (2k+1)\theta_K$ ($k = 0,1,\dots,K-1$), where I is on a decision boundary. Suppose $\arg(I) = (2k+1)\theta_K$, if the neuron state is either $e^{2ki\theta_K}$ or $e^{2(k+1)i\theta_K}$, the state remains. Otherwise, the new state is randomly selected from $e^{2ki\theta_K}$ and $e^{2(k+1)i\theta_K}$. Denote the set of neuron states by V. the activation function is as follows:

$$f(I) = \arg\max_{c \in V} \operatorname{Re}(\bar{c}I) \tag{6.22}$$

where $\operatorname{Re}(I)$ is the real part of I.

Let the connection weight from the j-th neuron to the k-th neuron be denoted by w_{kj}.

$$w_{kj} = \bar{w}_{jk} \tag{6.23}$$

where $\bar{w}$ is the complex conjugate of w. We denote the state of the j-th neuron by z_j. Then, the weighted sum input I_k to the k-th neuron is expressed as follows:

$$I_k = \sum_{j \neq k} w_{kj} z_j \tag{6.24}$$

The next state of neuron k is defined as $f(I_k)$ using asynchronous update.

Subsequently, we define the energy function. Let the state function be $z = (z_1, z_2, \ldots, z_N)$, where N is the number of neurons. Then, the energy E is defined as follows:

$$E = \frac{1}{2} \sum_k \sum_{j \neq k} \bar{z}_k w_{kj} z_j \qquad (6.25)$$

It is easy to prove $E = \bar{E}$. Therefore, E is a real number. When a neuron is updated, the energy never increases.

The complex-valued Hebbian learning rule is the simplest learning method. Let the p-th training pattern vector be denoted by $(c_1^p, c_2^p, \quad c_N^p,)$ $(p = 1, 2, \quad P)$, where P is the number of training patterns. Then, the complex-valued Hebbian learning rule produces the following connection weights,

$$w_{kj} = \sum_p c_k^p \bar{c}_j^p \qquad (6.26)$$

When the q-th training pattern is applied, the weighted sum input I_k to the k-th neuron is as follows:

$$I_k = \sum_{j \neq k} \sum_p c_k^p \bar{c}_j^p c_j^q = (N-1)c_k^q + \sum_{j \neq k} \sum_{p \neq q} c_k^p \bar{c}_j^p c_j^q \qquad (6.27)$$

If the absolute value of the second term of the right-hand side is small enough, we can conclude that the q-th training pattern is stable. The second term is referred to as the crosstalk term.

For simplicity, we consider the case of one training pattern $(c_1, c_2, \ldots, c_N)$. Then, a rotated pattern $(e^{i\theta} c_1, e^{i\theta} c_2, \ldots, e^{i\theta} c_N)$ is stable. When a rotated pattern is given, the weighted sum input I_k to the k-th neuron is as follows:

$$I_k = \sum_{j \neq k} c_k \bar{c}_j e^{i\theta} c_j = (N-1)e^{i\theta} c_k \qquad (6.28)$$

This is referred to as rotational invariance.

6.6   CONVOLUTIONAL NEURAL NETWORK

There are numerous variants of convolutional neural network (CNN) architectures, and their basic components are very similar. The CNN consists of three types of layers, namely convolutional, pooling, and fully connected layers. The convolutional layer aims to learn feature representations of the inputs and is composed of several convolution kernels which are used to compute different feature maps. Specifically, each neuron of a feature map is connected to a region of neighboring neurons in the previous layer. Such a neighborhood is referred to as the neuron's receptive field in the previous layer. The new feature map can be obtained by first convolving the input with a learned kernel and then applying an element-wise nonlinear activation function on the convolved results. Note that, to generate each feature map, the kernel is shared by all spatial locations of the input. The complete feature maps are obtained by using several different kernels. Mathematically, the feature value at location (i, j) in the k-th feature map of the l-th layer,

$$z_{i,j,k}^l = (w_k^l)^T x_{i,j}^l + b_k^l \tag{6.29}$$

where w_k^l and b_k^l are the weight vector and bias term of the k-th filter of the l-th layer respectively, and $x_{i,j}^l$ is the input patch centered at location (i, j) of the l-th layer. Note that the kernel w_k^l that generates the feature map is shared. Such a weight-sharing mechanism has several advantages such as it can reduce the model complexity and make the network easier to train. The activation function introduces nonlinearities to CNN, which are desirable for multilayer networks to detect nonlinear features. Let $a(\cdot)$ denote the nonlinear activation function. The activation value $a_{i,j,k}^l$ of the convolutional feature $z_{i,j,k}^l$ can be computed as

$$a_{i,j,k}^l = a(z_{i,j,k}^l) \tag{6.30}$$

Typical activation functions are sigmoid, tanh, and ReLU. The pooling layer aims to achieve shift-invariance by reducing the resolution of the feature maps. It is usually placed between two convolutional layers. Each feature map of a pooling layer is connected to its corresponding feature map of the preceding convolutional layer. Denoting the pooling function as pool($\cdot$), for each feature map $a_{m,n,k}^l$.

$$y_{i,j,k}^l = \text{pool}(a_{m,n,k}^l), \forall(m,n) \notin R_{ij} \tag{6.31}$$

where R_{ij} is a local neighborhood around the location (i, j). The typical pooling operations are average pooling and max pooling. The kernel in the first convolutional layer is designed to detect low-level features such as edges and curves, while the kernels in higher layers are learned to encode more abstract features. By stacking several convolutional and pooling layers, we could gradually extract higher-level feature representations.

After several convolutional and pooling layers, there may be one or more fully connected layers that aim to perform high-level reasoning. They take all neurons in the previous layer and connect them to every single neuron of the current layer to generate global semantic information. Note that a fully connected layer is not always necessary as it can be replaced by a 1*1 convolution layer.

The last layer of CNNs is an output layer. Let θ denotes all the parameters of a CNN (the weight vectors and bias terms). The optimum parameters for a specific task can be obtained by minimizing an appropriate loss function defined on that task.

Suppose we have N desired input–output relations $\{(x^{(n)}, y^{(n)}); n \in [1, \ldots, N]\}$, where $x^{(n)}$ is the n-th input data, $y^{(n)}$ is its corresponding target label, and $o^{(n)}$ is the output of CNN. The loss of CNN can be calculated as

$$L = \frac{1}{N} \sum_{n=1}^{N} l(\theta; y^{(n)}, o^{(n)}) \tag{6.32}$$

Training CNN is a problem of global optimization. By minimizing the loss function, we can find the best fitting set of parameters.

REFERENCES

1. Finol, D., Lu, Y., Mahadevan, V., et al. Deep convolutional neural networks for eigenvalue problems in mechanics. *International Journal for Numerical Methods in Engineering*, **2019**, 118(5), 258–275.
2. Kirchdoerfer, T, Ortiz, M. Data-driven computational mechanics. *Computer Methods in Applied Mechanics and Engineering*, **2016**, 304, 81–101.

Section II

Examples

7 Numerical Examples

7.1 IMPORTANCE SAMPLING

$$\mu = \int f(x)\pi(x)\,dx \qquad (7.1)$$

$f(x)$ is a measurable function and $\pi(x)$ is a probability density function.

The importance distribution function $g(x)$ is used to apply a change of measure

$$\mu = \int \frac{\pi(x)}{g(x)} f(x)g(x)\,dx \qquad (7.2)$$

If $\omega(X_i) = \dfrac{\pi(X_i)}{g(X_i)}$, $i = 1,\ldots,n$, then

$$\hat{\mu}_n^{IS} = \frac{1}{n}\sum_{i=1}^{n}\omega(X_i)f(X_i) \qquad (7.3)$$

When the example of Student-t distribution $T(v,\theta,\sigma^2)$ with density is performed as,

$$\pi(x) = \frac{\Gamma((v+1)/2)}{\sigma\sqrt{v\pi}\Gamma(v/2)}\left(1+\frac{(x-\theta)^2}{v\sigma^2}\right)^{-(v+1)/2} I_R(x) \quad (\theta = 0, \sigma = 1, v = 12) \quad (7.4)$$

The quantities of interest are supposed to be,

$$\begin{cases} f_1(x) = \left(\dfrac{\sin(x)}{x}\right)^5 I(x)_{(2.1,+\infty)} & (7.5) \\[3mm] f_2(x) = \sqrt{\left|\dfrac{x}{1-x}\right|} \\[3mm] f_3(x) = \dfrac{x^5}{1+(x-3)^2} I(x)_{[0,+\infty)} \end{cases}$$

When the following instrumental distributions are used as in following:

$$T(v^*,0,1) \text{ with } v^* < v, v^* = 7$$

DOI: 10.1201/9781003226628-9

$$N(0, v/(v-2))$$

$$C(0,1)$$

In addition, the Cauchy distribution $C(\alpha, \beta)$ has a density function

$$\pi(x) = \frac{1}{\pi\beta(1+((x-\alpha)/\beta)^2)} I(x) \tag{7.6}$$

The Monte Carlo simulation is performed to estimate $\hat{\mu}_n^{IS}$, plot 95% and 5% quantiles, and the mean of the estimator for $n = 1, ..., 50,000$ (Tables 7.1 and 7.2).

The fundamental issue in implementing the importance sampling simulation is the choice of the biased distribution which encourages the important regions of the input variables. Choosing or designing a good biased distribution is the key point of importance sampling. The rewards for a good distribution can be huge run-time savings; the penalty for a bad distribution can be longer run times than for a general Monte Carlo simulation (Figures 7.1–7.3).

In principle, the importance sampling ideas remain the same in these situations, but the design becomes much harder. A successful approach to combat this problem is essentially breaking down a simulation into several smaller, more sharply defined subproblems. Then importance sampling strategies are used to target each of the simpler subproblems.

To identify successful importance sampling approaches, it is useful to be able to quantify the run-time savings and computational efficiency. The performance measure commonly used is $\sigma_{MC}^2 / \sigma_{IS}^2$, and this can be interpreted as the speed-up factor by

TABLE 7.1
Time Cost of Different PDF by the Importance Sampling Method

	f1	f2	f3
Student-t	7.634	7.628	7.600
Normal	5.974	5.956	5.940
Cauchy	3.807	3.563	3.923
Exact	1.768	1.172	1.719

TABLE 7.2
Results Of Different PDF by the Importance Sampling Method

	f1	f2	f3
Student-t	7.705e-5	1.157	4.523
Normal	7.444e-5	1.167	4.659
Cauchy	7.984e-5	1.165	4.514
Exact	7.749e-5	1.164	4.708

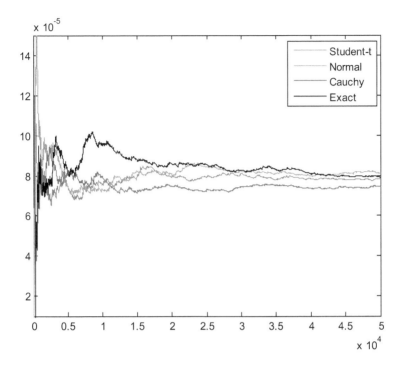

FIGURE 7.1   Convergence history of different PDF in function $f1$.

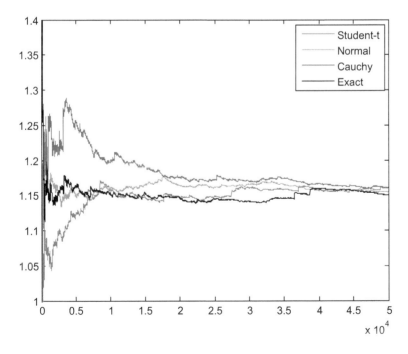

FIGURE 7.2   Convergence history of different PDF in function $f2$.

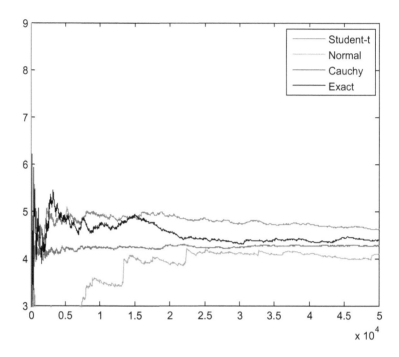

FIGURE 7.3   Convergence history of different PDF in function $f3$.

which the importance sampling estimator achieves the same precision as the Monte Carlo estimator. This has to be computed empirically since the estimator variances are not likely to be analytically possible when their mean is intractable. Other useful concepts in quantifying an importance sampling estimator are the variance bounds and the notion of asymptotic efficiency.

7.2   ORTHOGONAL POLYNOMIAL

In the one-dimensional case, the random response u is expanded by the orthogonal polynomials in ξ, which has a known probability distribution such as unit normal, $N[0,1]$. If u is a function of a normally distributed random variable x, which has the known mean μ_x and variance σ_x^2, ξ is a normalized variable:

$$\xi = \frac{x - \mu_x}{\sigma_x} \tag{7.7}$$

Generally, the one-dimensional Hermite polynomials are defined by

$$\Psi_n(\xi) = (-1)^n \frac{\varphi^n(\xi)}{\varphi(\xi)} \tag{7.8}$$

where $\varphi^n(\xi)$ is the n th derivative of the normal density function, $\varphi(\xi) = 1/\sqrt{2\pi}e^{-\xi^2/2}$. This is simply the single-variable version.

$$\{\Psi_i\} = \left\{1, \ \xi, \ \xi^2 - 1, \ \xi^3 - 3\xi, \ \xi^4 - 6\xi^2 + 3, \ \xi^5 - 10\xi^3 + 15\xi, \ \cdots\right\} \quad (7.9)$$

Thus, a second-order, 2D PCE is given by

$$u(\theta) = b_0 + b_1\xi_1(\theta) + b_2\xi_2(\theta) + b_3(\xi_1^2(\theta) - 1) + b_4\xi_1(\theta)\xi_2(\theta) + b_5(\xi_2^2(\theta) - 1) \quad (7.10)$$

where $\xi_1(\theta)$ and $\xi_2(\theta)$ are two independent random variables.

Suppose the random variable x that is normally/nonnormally distributed. This random variable x can be approximated by the first four terms of the PCE as follows:

$$x \approx z(\xi) = b_0 + b_1\xi + b_2(\xi^2 - 1) + b_3(\xi^3 - 3\xi) \quad (7.11)$$

Calculate the first four central moments of z in terms of the coefficients b_i.

The standard normal random variable ξ and orthogonal polynomials Ψ_i satisfy

$$\Psi_0 = 1, E[\Psi_i] = 0 \quad (7.12)$$

$$E[\Psi_i\Psi_j] = E[\Psi_i^2]\delta_{ij}, \delta_{ij} = \begin{cases} 1, & i = j \\ 0, & i \neq j \end{cases} \quad (7.13)$$

where δ_{ij} is the Kronecker delta.

Suppose the first four moments of a random variable, x, are given by $m_x^1 = 2$, $m_x^2 = 3.2$, $m_x^3 = 7.5$, $m_x^4 = 45$. Estimate the coefficients b_i of z by using the least-squares criterion (Figure 7.4):

$$\text{Minimize} \sum_{j=1}^{4} f_j^2(b_i) \quad (i = 1, 2, 3)$$

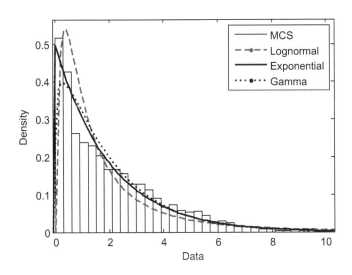

FIGURE 7.4 In the Monte Carlo simulation, 5000 Gaussian sampling points are generated.

where

$$f_1(b_i) = m_z^1 - m_x^1 = 0$$
$$f_2(b_i) = m_z^2(b_i) - m_x^2$$
$$f_3(b_i) = m_z^3(b_i) - m_x^3$$
$$f_4(b_i) = m_z^4(b_i) - m_x^4$$

Then perform optimization progress to have a solution of coefficients b_i and make sure to minimize $\displaystyle\sum_{j=1}^{4} f_j^2(b_i)$

$$E\left[\xi^k\right] = 0 \quad \forall k \text{ odd}$$

$$E\left[\xi^0\right] = 1, E\left[\xi^2\right] = 1, E\left[\xi^4\right] = 3, E\left[\xi^6\right] = 15...$$

$$m_z = b_0$$

$$m_z^2 = E\left[(z - b_0)^2\right] = b_1^2 + 2b_2^2 + 6b_3^2$$

$$m_z^3 = E\left[(z - b_0)^3\right] = 6b_1^2 b_2 + 8b_2^3 + 36b_1 b_2 b_3 + 108 b_2 b_3^2$$

$$m_z^4 = E\left[(z - b_0)^4\right] = 3b_1^4 + 60b_2^4 + 3384 b_3^4 + 24 b_1^3 b_3 + 60 b_1^2 b_2^2 + 252 b_1^2 b_3^2$$

$$+ 576 b_1 b_2^2 b_3 + 1296 b_1 b_3^3 + 2232 b_2^2 b_3^2$$

Then $b_0 = 2$, $b_1 = -1.545$, $b_2 = 0.628$, $b_3 = 0.141$

$$x \approx z(\xi) = 2 - 1.545\xi + 0.628(\xi^2 - 1) + 0.141(\xi^3 - 3\xi)$$

$$m^1 = 2.0049,\ m^2 = 3.2364,\ m^3 = 7.0124,\ m^4 = 43.2743$$

7.3 GRAM–CHARLIER SERIES

Specify the first seven coefficients of the Gram–Charlier series.

$$b_0 = \int_{-\infty}^{+\infty} f(x)\, dx, \tag{7.14}$$

$$b_1 = -\int_{-\infty}^{+\infty} f(x) x\, dx \tag{7.15}$$

$$b_2 = \frac{1}{2} \int_{-\infty}^{+\infty} f(x)(x^2 - 1)\, dx \tag{7.16}$$

$$b_3 = -\frac{1}{6} \int_{-\infty}^{+\infty} f(x)(x^3 - 3x)\, dx, \tag{7.17}$$

$$b_4 = \frac{1}{24} \int_{-\infty}^{+\infty} f(x)(x^4 - 6x^2 + 3)\, dx, \tag{7.18}$$

$$b_5 = -\frac{1}{120} \int_{-\infty}^{+\infty} f(x)(x^5 - 10x^3 + 15x)\, dx, \tag{7.19}$$

$$b_6 = -\frac{1}{720} \int_{-\infty}^{+\infty} f(x)(x^6 - 15x^4 + 45x^2 - 15)\, dx, \tag{7.20}$$

The n th order central moment can be given by

$$m_x^n = E\big[(X - \mu_x)^n\big] = \int_{-\infty}^{+\infty} (X - \mu_x)^n f_X(x)\, dx \tag{7.21}$$

Let the first moment be zero $\mu_x = m^1 = 0$.
Then,

$$b_0 = 1,\; b_1 = 0,\; b_2 = \frac{1}{2}(m^2 - 1),\; b_3 = -\frac{1}{6}(m^3 - 3m^1)$$

$$b_4 = \frac{1}{24}(m^4 - 6m^2 + 3),\; b_5 = -\frac{1}{120}(m^5 - 10m^3 + 15m^1),$$

$$b_6 = \frac{1}{720}(m^6 - 15m^4 + 45m^2 - 15)$$

The coefficients of the Gram–Charlier series can be expressed by Hermite polynomials in terms of central moments.

Suppose a target covariance matrix is given by

$$[C] = \begin{bmatrix} 1 & 0.63 & 0.75 & 0.92 \\ 0.63 & 1 & 0.96 & 0.84 \\ 0.75 & 0.96 & 1 & 0.72 \\ 0.92 & 0.84 & 0.72 & 1 \end{bmatrix}$$

Generate the correlated random variables (normal distribution) and compare the sample covariance matrix $\big[\tilde{C}\big]$

$$P = \begin{bmatrix} 0.4320 & -0.4475 & 0.6166 & 0.4827 \\ 0.5744 & 0.3822 & -0.5195 & 0.5040 \\ -0.4851 & -0.5522 & -0.4547 & 0.5030 \\ -0.4981 & 0.5905 & 0.3784 & 0.5099 \end{bmatrix}$$

$$\Lambda = \begin{bmatrix} -0.0653 & 0 & 0 & 0 \\ 0 & 0.1733 & 0 & 0 \\ 0 & 0 & 0.4807 & 0 \\ 0 & 0 & 0 & 3.4114 \end{bmatrix}$$

$$[A] = [P][\Lambda]^{1/2} = \begin{bmatrix} 0 & -0.1863 & 0.4275 & 0.8915 \\ 0 & 0.1591 & -0.3602 & 0.9309 \\ 0 & -0.2299 & -0.3152 & 0.9291 \\ 0 & 0.2458 & 0.2623 & 0.9418 \end{bmatrix}$$

$$\left[\tilde{C}_{500}\right] = \begin{bmatrix} 1.1109 & 0.7590 & 0.8411 & 0.9690 \\ 0.7590 & 1.0681 & 0.9923 & 0.8565 \\ 0.8411 & 0.9923 & 1.1375 & 0.8227 \\ 0.9690 & 0.8565 & 0.8227 & 1.0850 \end{bmatrix}$$

7.4   KRIGING SURROGATE MODEL

A surrogate model is like a black box, where the mathematical relationship between the input variables and output results is expressed by approximated implicit methods. Usually, the surrogate model can be classified into two groups: local and global models. The response surface method is a typical local model and can be written in the polynomial series as [1–4]

$$F(\beta : x) = \beta_0 + \sum_{i=1}^{n} \beta_i x_i \tag{7.22}$$

$$F(\beta : x) = \beta_0 + \sum_{i=1}^{n} \beta_i x_i + \sum_{i=1}^{n} \sum_{j=1}^{n} \beta_{ij} x_i x_j \tag{7.23}$$

Equations (7.22) and (7.23) are the local first- and second-order response surface models, respectively. $F(\beta : x)$ is a deterministic regression model, and β_i is the corresponding regression coefficient.

In addition, a global surrogate model generally has global searching space. The Kriging model fits a spatial correlation function as shown below:

$$G(x) = F(\beta : x) + z(x) \tag{7.24}$$

where $z(x)$ is assumed to have zero mean and covariance.

In the first-order linear Kriging surrogate model, the predictor can be expressed as

$$\hat{y}(x) = c^T Y \tag{7.25}$$

with $c = c(x) \in R^m$. The relative error is

$$\hat{y}(x) - y(x) = c^T Y - y(x)$$

$$= c^T (F\beta + Z) - (f(x)^T \beta + z)$$

$$= c^T Z - z + (F^T c - f(x))^T \beta, \tag{7.26}$$

To make sure the predictor is unbiased, $F^T c$ is equal to $f(x)$. The mean squared error of the predictor can be written as

$$\varphi(x) = E\left[(\hat{y}(x) - y(x))^2\right]$$

$$= E\left[(c^T Z - z)^2\right]$$

$$= \sigma^2(1 + c^T Rc - 2c^T r) \tag{7.27}$$

The objective of the optimization program is to minimize φ by satisfying the following constraint.

$$L(c, \lambda) = \sigma^2(1 + c^T Rc - 2c^T r) - \lambda^T (F^T c - f) \tag{7.28}$$

The computation of the gradient of the constraint function with respect to c can be expressed as

$$L_c'(c, \lambda) = 2\sigma^2(Rc - r) - F\lambda \tag{7.29}$$

Suppose $\tilde{\lambda} = -\dfrac{\lambda}{2\sigma^2}$, the system equations are

$$\begin{bmatrix} R & F \\ F^T & 0 \end{bmatrix} \begin{bmatrix} c \\ \tilde{\lambda} \end{bmatrix} = \begin{bmatrix} r \\ f \end{bmatrix} \tag{7.30}$$

The solution can be obtained,

$$\tilde{\lambda} = (F^T R^{-1} F)^{-1}(F^T R^{-1} r - f)$$

$$c = R^{-1}(r - F\tilde{\lambda}) \tag{7.31}$$

where R^{-1} is the inverse matrix of the correlation matrix R.

$$\hat{y}(x) = (r - F\tilde{\lambda})^T R^{-1} Y$$

$$- r^T R^{-1} Y - (F^T R^{-1} r - f)^T (F^T R^{-1} F)^{-1} F^T R^{-1} Y \tag{7.32}$$

The generalized least squares solution is

$$\beta^* = (F^T R^{-1} F)^{-1} F^T R^{-1} Y \tag{7.33}$$

Substitute β^* into the predictor of the Kriging surrogate model,

$$\hat{y}(x) = r^T R^{-1} Y - (F^T R^{-1} r - f)^T \beta^*$$

$$= f^T \beta^* + r^T R^{-1}(Y - F\beta^*)$$

$$= f(x)^T \beta^* + r(x)^T \gamma^* \tag{7.34}$$

Besides, the corresponding maximum likelihood estimate of the variance is written as

$$\sigma^2 = \frac{1}{m}(Y - F\beta^*)^T(Y - F\beta^*) \tag{7.35}$$

If the errors are uncorrelated and have different variances, $E[e_i e_i] = \sigma_i^2$ and $E[e_i e_j] = 0$ for $i \neq j$. It is logical to find that R is the diagonal matrix,

$$R = \text{diag}\left(\frac{\sigma_1^2}{\sigma^2}, \ldots, \frac{\sigma_m^2}{\sigma^2}\right) \tag{7.36}$$

Besides, the weight matrix W is given by

$$W = \text{diag}\left(\frac{\sigma}{\sigma_1}, \ldots, \frac{\sigma}{\sigma_m}\right) \Leftrightarrow W^2 = R^{-1} \tag{7.37}$$

Therefore,

$$\tilde{Y} = WY = WF\beta + \tilde{e} \tag{7.38}$$

The below equations are satisfied:

$$E[\tilde{e}] = 0, \quad E[\tilde{e}\tilde{e}^T] = WE[ee^T]W^T = \sigma^2 I \tag{7.39}$$

Replacing F and Y by the weighted function, the results can be depicted as

$$(F^T W^2 F)\beta^* = F^T W^2 Y,$$

$$\sigma^2 = \frac{1}{m}(Y - F\beta^*)^T W^2 (Y - F\beta^*) \tag{7.40}$$

7.4.1 NUMERICAL ISSUES OF THE KRIGING MODEL

The Kriging model is widely applied in engineering and related fields, based on its solid mathematical foundation and convenience in implementation. However, as mentioned in the introduction section, the Kriging model inevitably confronts some numerical issues, which can be concluded into two categories:

7.4.1.1 Issue 1: Requirement of Sufficient Effective Samples

To have more precise surrogate models, it is crucial to offer sufficient effective samples. The requirement for the database is not only to provide a large number of samples, but also to ensure the effectiveness and typical representation of each sample. Nevertheless, a large number of samples cause high computation expenses and bring difficulties in the fitting and prediction process. In addition, the selection process is necessary to proceed with choosing the typical and representative samples. Simple examples are presented to discuss the issue.

Examples:

$$\begin{cases} f_1 = \cos x \Big/ \exp\left(\dfrac{x}{10}\right) \\ f_2 = x^4 + 10x^3 + 11x^2 - 94x - 168 \\ f_3 = \sin x \end{cases} \tag{7.41}$$

Figure 7.5 shows that the erratic behavior of the Kriging model and the results are identical to the problem in issue 1. Although the model passed through all sample points, the prediction at arbitrary locations was erratic and inaccurate. This situation frequently happens if a small number of sample points are equally distributed.

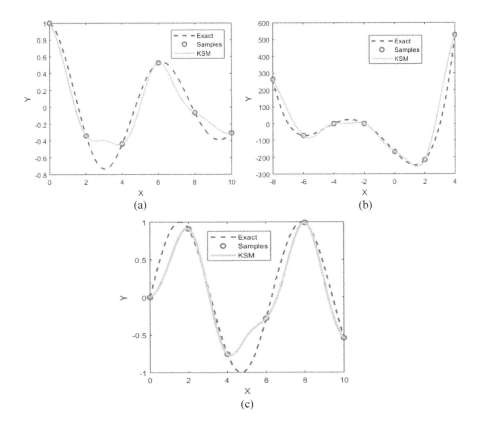

FIGURE 7.5 Examples of the Kriging models without sufficient effective samples (a–c are for Examples 1–3, respectively).

7.4.1.2   Issue 2: Effects of the Correlation Matrix

The correlation functions can be divided into two groups: a set of functions that have parabolic behavior near the origin (the Gauss's, cubic, spline function) and another group of functions (the exponential, linear, spherical) that have linear behavior near the origin. According to the last parameter, the general exponential function has two shapes: the giving Gauss and exponential functions, respectively.

Assuming a Gaussian process, the optimal coefficients θ^* of the correlation function solves

$$\min_{\theta}\left\{\psi(\theta) \equiv |R|^{\frac{1}{m}}\sigma^2\right\} \tag{7.42}$$

where $|R|$ is the determinant of R. This definition of θ^* corresponds to maximum likelihood estimation.

Apart from the previous issue, unstable behavior with a sufficient number of sample points is also observed. The correlation matrix should be always positive definite and seldom singular. Otherwise, the correlation matrix causes the Kriging model unstable and erratic behavior. The Kriging model is largely influenced by the condition of the correlation matrix, as presented in Figure 7.6. A large fluctuation illustrates that besides the sufficient effective sample database, an appropriate correlation matrix is another important factor, which directly affects the accuracy of the Kriging model predictor. Moreover, the results in Figure 7.6 also show that large variances in the Kriging model prediction with an infeasible correlation matrix are found. The linear, spherical, cubic spline correlation matrix caused unstable and erratic behavior in the mentioned simple examples. In contrast, the Gaussian correlation matrix has more accurate and satisfying prediction results by the Kriging model than others.

7.4.2   Subset Simulation

Usually, the state function F is defined as the subregion in the **x**-space. The space exceeds the performance function $g(\mathbf{x})$ and is below a certain threshold value b, as follows [1–2]:

$$F = \left\{\mathbf{x} : g(\mathbf{x}) < b\right\} \tag{7.43}$$

where **x** is the input random vector and $g(\mathbf{x})$ can be an implicit or complicated nonlinear function. The essential idea of SS is to divide the space of input variables into specific subset domains. Therefore,

$$F_j = \left\{\mathbf{x} : g(\mathbf{x}) < b_j\right\}, \quad j = 1,\ldots,m \qquad \cdot \tag{7.44}$$

where m is the total number of intermediate events, and b_j represents the corresponding threshold values.

The state function can be expressed as

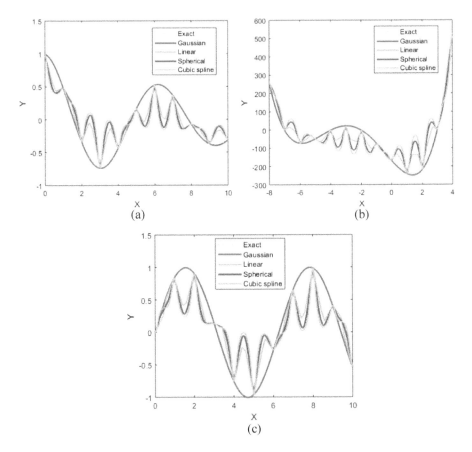

FIGURE 7.6 Examples of the Kriging models with different correlation matrices (a–c are for Examples 1–3, respectively).

$$P_F = P(F) = P(F_1)\prod_{j=1}^{m-1} P(F_j|F_{j-1}) \qquad (7.45)$$

The event F is converted to be a series of conditional events $F_j|F_{j-1}$.

The subsequent conditional probabilities $P_j = P(F_j | F_{j-1})$ require the samples conditioning on F_{j-1} with implicit conditional PDF

$$f(\mathbf{x}|F_{j-1}) = f(\mathbf{x})I_{F_{j-1}}(\cdot)/P(F_{j-1}) \qquad (7.46)$$

An effective procedure based on the Markov Chain Monte Carlo simulation can be used to provide the required conditional samples $\{x_i\}$ and then the conditional probabilities for $P(F_j|F_{j-1})$

$$P_j = P(F_j|F_{j-1}) \approx \frac{1}{N}\sum_{i=1}^{N} I_{F_{j-1}}(g(\mathbf{x}_i)) \qquad (7.47)$$

The state function of the target event is rewritten as

$$P_F = \prod_{j=1}^{m} P_j \tag{7.48}$$

7.4.3 KRIGING SURROGATE MODEL WITH SUBSET SIMULATION

7.4.3.1 The Six-Hump Camel-Back Function

Consider the problem of creating a Kriging model for the six-hump camel-back function, which is a typical nonmonotonic function. The exact contour picture is plotted in Figure 7.7. The mathematical function is presented as

$$h(x) = 4x_1^2 - 2.1x_1^4 + x_1^6/3 + x_1x_2 - 4x_2^2 + 4x_2^4 \quad \mathbf{x} \in [-3,3] \tag{7.49}$$

The relative error is computed by

$$e = \frac{\sum_{i}^{n} (\hat{y}_i - y_i)^2}{n} \tag{7.50}$$

where $\hat{y}$ is the prediction result of the Kriging surrogate model based on the subset simulation, and y denotes the exact result of the six-hump camel-back function. From

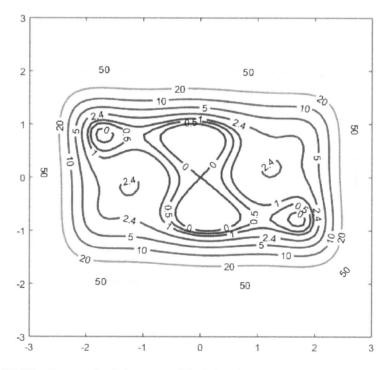

FIGURE 7.7   Contour of a six-hump camel-back function.

the prediction of the Kriging surrogate model based on the subset simulation, the relative error is demonstrated in Figure 7.8.

For the subset selection of the six-hump camel-back function, the sampling points are picked up according to the symmetry of the geometrical center (0, 0). The distribution of input variables in the sampling process of subset simulation is uniform in the corresponding interval ranges. In order to compare the efficiency improvement of the Kriging surrogate model based on subset simulation, the first- and second-order polynomial regression (Poly-1, Poly-2) and the ordinary Kriging surrogate model (KSM) are performed and the results are demonstrated in Figure 7.8.

In Figure 7.8, the relative errors of prediction results from the first- and second-order polynomials decrease at first with the increase of the number of sampling points, while the accuracy improvement of the first- and second-order polynomials is not obvious according to the increase of the number of sampling points, especially when the amount of the sampling points exceeds 40. In addition, the ordinary KSM has evident fluctuation in the accuracy of the prediction results. Even though the relative errors of prediction results from the ordinary KSM are reduced more sharply than those of first- and second-order polynomials, the fluctuation in the accuracy of results makes the ordinary KSM unstable. On the contrary with the polynomials and ordinary KSM, with the increase of the number of sampling points in the KSM based on the subset simulation, the accuracy of prediction results dramatically improves and quickly reaches a satisfying level. Therefore, the feasibility of

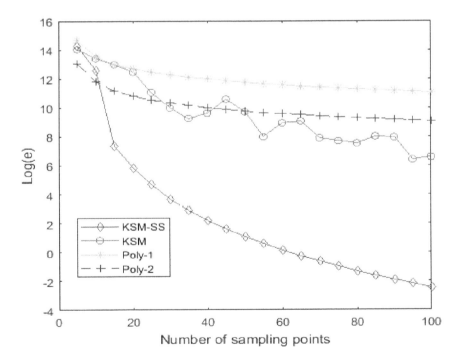

FIGURE 7.8 Prediction accuracy of the Kriging surrogate model based on the subset simulation (Poly-1 and Poly-2 are first- and second-order polynomial regression, respectively).

efficiency improvement by the KSM based on the subset simulation is proved in the explicit problem of the six-hump camel-back function.

7.4.3.2 Vibration Analysis of a Wing Structure

In the fields of engineering, the implicit expression problems often involve aleatory uncertainty and complicated nonlinearity. The KSM is an effective alternative method for the representation of the implicit function or relationship. In the vibration analysis of wing structures, the parameters corresponding to the geometrical (length of wing (L), parameters describing the airfoil (S, D)) and material properties (Young's modulus (E), Poisson ratio (R), and physical density (P)) are input variables, and the resonant frequencies of different order vibration modes are output results. The vibration modes are calculated by performing the finite element in the deterministic model to confirm the original finite element model of the wing structure (Figure 7.9).

To create the KSM based on the subset simulation for the vibration analysis of wing structures, a flowchart is presented in Figure 7.10. The comprehensive

FIGURE 7.9   Vibration mode of the deterministic finite element model.

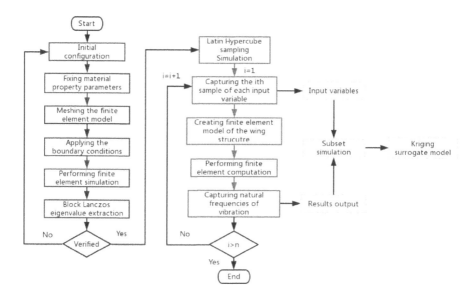

FIGURE 7.10   Flowchart of the Kriging surrogate model implementation.

procedure includes three entire components, and each component is marked with three different colors, blue, red, and green. The processes in the program marked with blue color represent the deterministic finite element model for wing structures. After the validation of the deterministic finite element model, the Latin hypercube sampling simulation is performed to define the input variables, which are corresponding to the geometrical and material properties of wing structures. On the other hand, the red and green colors in the flowchart signify the repeat of performing finite element simulation for vibration analysis of wing structures and the creation of the subsets for the KSM, respectively. The finite element program is performed repeatedly until sufficient times are completed. The results in the stochastic finite element are captured and transferred into the procedure of subsets forming. Even though the relationship between the response of wing structures and the parameters related to geometrical and material properties in vibration is implicit and nonlinear in the expression, by the KSM based on the subset simulation, an effective surrogate model can work well.

Different from the ordinary KSM, the input variables in the wing structures are not only sampled randomly by Latin hypercube sampling simulation following uniform distribution, but also captured and selected from the original databases of Latin hypercube sampling simulation for subsets forming as in Figure 7.11. According to the corresponding resonant frequencies of wing structures, the original database of sampling points of input variables is rearranged in order. The linear spaces are applied to divide the interval range of resonant frequencies into certain amounts of subsets for the computation of the probability density distribution. Corresponding to the subsets of resonant frequencies, the subsets of input variables are created. In addition, the number of sampling points of each input variable from the forming subsets is determined by the probability density distribution of resonant frequencies. The selected sampling points of each input variable from the subsets are transferred to the KSM. The efficiency improvement of the KSM based on the subset simulation in the implicit expression of wing structures is discussed in the following.

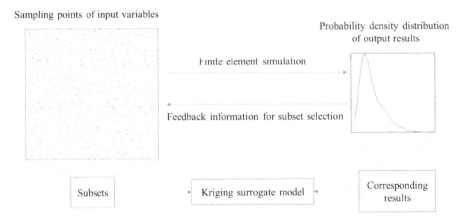

FIGURE 7.11 Subsets forming based on the database of Latin hypercube sampling.

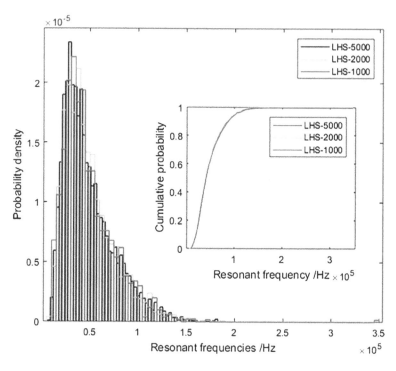

FIGURE 7.12   Convergence of Latin hypercube sampling simulation in the original database.

The results of probability density distribution and accumulative probability of resonant frequencies are presented in Figure 7.12. The histograms of probability density distribution and the curves of the accumulative probability of resonant frequencies of wing structures demonstrate that the Latin hypercube sampling-based finite element model has a good convergence in the number of sampling points of input variables. In the Latin hypercube method, 1000 sampling points of each input variable are sufficient to provide satisfied accurate results. Thus, the original database of the Latin hypercube sampling-based finite element model for the vibration analysis of wing structures is believable and appropriate for the KSM based on the subset simulation.

The efficiency improvement of the KSM based on the subset simulation is tested by the relative errors of the prediction. In Figure 7.13, different examples are predicted by the KSM. By comparing with the accurate results of finite elements for wing structures, the relative errors are computed. The relative errors of the prediction results of the KSM based on the subset simulation are smaller than that of the ordinary KSM, especially when the number of sampling points in each input variable is small. The relative errors of prediction results decline dramatically when the amount of the sampling points provided by subset simulation increases. The KSM based on the subsets simulation provides accurate prediction results with fewer sampling points in input variables, which not only enhances the accuracy of the KSM but also reduces the computational costs.

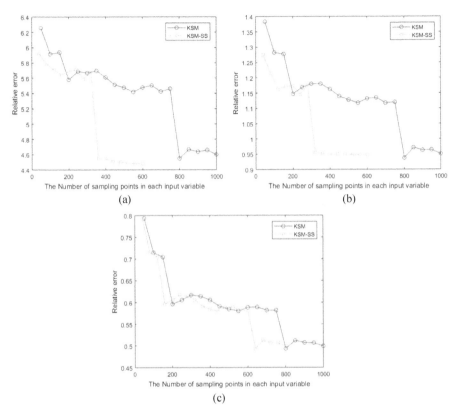

(a)   (b)

(c)

FIGURE 7.13 Comparison of the accuracy of an ordinary Kriging surrogate model and a Kriging surrogate model based on the subset simulation (a–c for the prediction of 100, 500, and 1000 examples, respectively).

Additionally, besides the subset simulation, the relative errors are decreased according to the number of prediction examples. When the amount of prediction examples is increased, the advantages of the KSM based on the subset simulation are kept in the prediction accuracy. Therefore, the KSM based on the subset simulation is more competitive in computational costs and accuracy insurance of a small number of prediction examples.

The probability density distributions of the ordinary KSM and the KSM based on the subset simulation are compared with that of precise results. In Figure 7.14a, the KSM based on the subset simulation offers more approximated probability density distribution of resonant frequencies for vibration analysis of wing structures to the precise results than that of the ordinary KSM. In the peak and drag components of probability density distribution, the ordinary KSM has deviation to the precise results, but the results of the KSM based on the subset simulation are closer to that of the precise results. In Figure 7.14b, when the number of prediction examples is as large as 500, the probability density distributions of both the ordinary KSM and the KSM based on the subset simulation demonstrate satisfying results. Thus, the

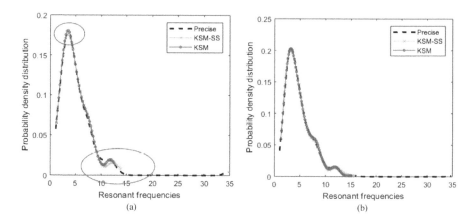

FIGURE 7.14   Probability density distribution of predicted results from the Kriging surrogate model (a and b are for 100 and 500 prediction examples, respectively).

KSM based on subset simulation not only reduces the computational cost with the insurance of accuracy level for the sampling database but also has the competence to provide more reliable prediction results than the ordinary KSM.

REFERENCES

1. Chu, L., Shi, J., Souza de Cursi, E., et al. Efficiency improvement of Kriging surrogate model by subset simulation in implicit expression problems. *Computational and Applied Mathematics*, **2020**, 39(2), 119.
2. Chu, L., Shi, J., Souza de Cursi, E. Kriging surrogate model for resonance frequency analysis of dental implants by Latin Hypercube based finite element method. *Applied Bionics and Biomechanics*, **2019**, 3768695, 1–14.
3. Shi, J., Chu, L., Braun, R. A Kriging surrogate model for uncertainty analysis of graphene based on a finite element method. *International Journal of Molecular Sciences*, **2019**, 20(9), 2355.
4. Chu, L., Shi, J., Souza de Cursi, E. A Kriging surrogate model for the interference reduction in the settlement surveillance sensors of steel transmission towers. *Applied Sciences*, **2019**, 9(16), 3343.

8 Monte Carlo-Based Finite Element Method

8.1 INTRODUCTION

Graphene sheets are promising nanomaterials with extraordinary properties for wide applications. A great amount of researches and experiments have been conducted to explore the properties of single-layer graphene sheets. However, vacancy defects are unavoidable existence, which crucially affects the mechanical properties of graphene sheets [1–4]. The large deviation in simulations and experiments is attributed to the presence and uncertainty of defects in the nanotube structure [5,6]. Therefore, attempts and struggles in the vacancy defects research of graphene sheets are very necessary, by which the deviation or fluctuation in simulations and experiments can be reasonably explained.

The main difficulty of the graphene mechanical behavior analysis is its small size [7,8]. It is hard to make accurate measurements in physical experiments at a nanometer scale, while analytical and numerical methods are powerful and alternative methods in the investigation. The solid atomic-based methods are promising in mechanical and electronic simulation, especially molecular dynamics simulation [9] and tight-binding molecular dynamics [10]; besides, the density functional theory [11] is another powerful supplement. On the other hand, the size-dependent continuum theories are also applied in the prediction and evaluation of graphene sheets, such as the non-local elastic theory [12], modified couple stress theory [13], and strain gradient theory [14]. However, when the amount of atoms is large in a system, atomistic-based methods cost large computational expenses; while in continuum theories, the defects or aleatory uncertainties in nanostructures are ignored or performed in complicated ways. It is necessary to develop an appropriate stochastic finite element method to propagate the randomly distributed vacancy defects in the deterministic finite element method.

In this challenging field, effective attempts and efforts have been made to analyze the influence of different defects. Banhart et al. [15] pointed out that a small number of defects in atomic structures of nanomaterials can profoundly influence the mechanical and electronic properties of graphene sheets. In addition, molecular dynamics [16] is used to analyze the influence of vacancy defects on the parameters of material properties. By the same method, the Stone–Wales defects of carbon nanotubes under a load of tension were evaluated [17,18]. However, the appropriate quantification of the influence of vacancy defects of graphene sheets in dynamic nonlinear processes, such as free vibration and buckling, still faces unexplained problems [19]. The stochastic and unpredictable properties of vacancy defects and their sophisticated effects on the mechanical response of graphene sheets are unsolved problems, which deserve more attention.

DOI: 10.1201/9781003226628-10

As one of the sophisticated sampling methods, the Monte Carlo simulation (MCS) is compatible with program design and has been widely adopted in the fields of research and engineering [20–23]. When the sampling space is large enough, the MCS can achieve acceptable accuracy in numerical results. It is common to settle the results of the MCS as comparison criteria or exact results [24,25]. In addition, the combination of the MCS and finite element analysis is feasible and convenient to perform. A combination of the MCS and the finite element method (FEM) forms a stochastic finite element method called Monte Carlo-based finite element method (MC-FEM), by which the randomly dispersed vacancy defects can be successfully propagated and simulated in graphene sheets.

8.2 GRAPHENE MATERIAL DESCRIPTION

The validation of modified Morse potential has been proved in previous studies to predict the mechanical properties of carbon nanotubes. The potential was successfully employed in the nonlinear response simulation of nanomaterials under tensile and torsional external forces [26]. The effects of defects on Young's modulus of nanotubes were investigated using the modified Morse potential [27]. In comparative studies, the modified Morse potential provides more precise prediction results than the reactive empirical bond order potential [28,29]. As stated in the previous work, the potential energy of the entire graphene sheets is written as [30]

$$E = E_s + E_a$$

$$E_s = D\left[\left(1 - e^{-\beta(r-r_0)}\right)^2 - 1\right]$$

$$E_a = \frac{1}{2}k_\theta(\theta - \theta_0)^2\left[1 + k_s(\theta - \theta_0)^4\right] \tag{8.1}$$

where E_s is the bond energy owing to the bond stretching; E_a is the bond energy corresponding to the angle-bending; r is the bond length; and θ is the angle of the adjacent bond. The parameters of the potential are [30]:

$$r_0 = 1.42\times10^{-10} \text{ m}, D = 6.031\times10^{-19} \text{ Nm}, \beta = 2.625\times10^{10} \text{ m}^{-1}$$

$$\theta_0 = 2.094 \text{ rad}, k_\theta = 0.9\times10^{-18} \text{ Nm / rad}^2, k_s = 0.754 \text{ rad}^{-4}$$

Based on the above expression, the diameter d, Young's modulus E, and shear modulus G of the beam elements in the honeycomb lattice of graphene sheets can be calculated according to

$$\begin{cases} d = 4\sqrt{\dfrac{k_\theta}{k_r}} \\[2mm] E = \dfrac{k_r^2 L}{4\pi k_\theta} \\[2mm] G = \dfrac{k_r^2 k_\tau L}{8\pi k_\theta^2} \end{cases} \tag{8.2}$$

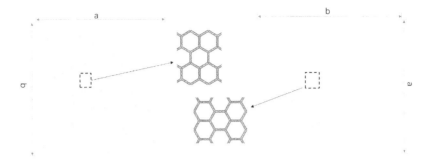

FIGURE 8.1 Schematic of monolayer graphene (the zigzag and armchair types).

where k_r, k_θ, and k_τ are the bond sketching, bond bending, and torsional resistance force constants, respectively. Figure 8.1 illustrates a schematic diagram of monolayer graphene in two dimensions.

A circular solid cross-sectional area is assumed in BEAM188, and the wall thickness corresponds to the section diameter. It is desired to compute the natural frequencies of graphene sheets in different vibration modes. Figure 8.1 illustrates a schematic of monolayer graphene in two dimensions. The zigzag and armchair types are both included.

The numerical simulation model of graphene sheet was created by ANSYS parameter design language. Carbon atoms in graphene are bonded together with covalent bonds forming a hexagonal 2D lattice. The displacement of individual atoms under an external force is constrained by the bonds. Therefore, the total deformation of graphene is the result of the interactions between the bonds.

For the modeling of the C–C bonds, the 3D elastic BEAM188 element was used. BEAM188 is suitable for analyzing slender to moderately stubby/thick beam structures. The element is based on the Timoshenko beam theory which includes first-order shear-deformation effects. The element is a linear, quadratic, or cubic two-node beam element in 3D. For each node, it has six degrees of freedom, which include translations in the x, y, and z directions and rotations around the x, y, and z directions. In the truss FEM, there are 6226 trusses, 4212 points, and 16,664 nodes created.

8.3 MONTE CARLO-BASED FINITE ELEMENT METHOD

Let ξ be a random variable for which the mathematical expectation of $E(\xi) = I$ exists. It is formally defined as

$$E(\xi) = \begin{cases} \int_{-\infty}^{\infty} \xi p(\xi) d\xi & \text{where } \int_{-\infty}^{\infty} p(x)dx = 1 & \text{when } \xi \text{ is a continuous r.v.} \\ \Sigma_\xi \, \xi p(\xi) & \text{where } \Sigma_x p(x)x = 1 & \text{when } \xi \text{ is a discrete r.v} \end{cases} \tag{8.3}$$

The nonnegative function $p(x)$ (continuous or discrete) is called the probability density function. To approximate the variable I, a computation of the arithmetic mean must be carried out,

$$\bar{\xi}_N = \frac{1}{N} \sum_{i=1}^{N} \xi_i \tag{8.4}$$

For a sequence of uniformly distributed independent random variables, the arithmetic mean of these variables converges to the mathematical expectation:

$$\xi_N \xrightarrow{p} I \quad \text{as} \quad N \to \infty$$

The sequence of the random variables $\eta_1, \eta_2, \ldots, \eta_N, \ldots$ converges to the constant c if, for every $h > 0$, it follows that

$$\lim_{N \to \infty} P\{|\eta_N - c| \geq h\} = 0$$

Thus, when N is sufficiently large $\bar{\xi}_N \approx I$.

Suppose that the random variable ξ has a finite variance, the error of the algorithm can be estimated as

$$D(\xi) = E[\xi - E(\xi)]^2 = E(\xi^2) - [E(\xi)]^2 \tag{8.5}$$

Crude Monte Carlo is the simplest possible approach for solving multidimensional integrals. This approach applies the definition of the mathematical expectation.

Let Ω be an arbitrary domain and $x \in \Omega \subset R^d$ be a d-dimensional vector. We consider the problem of the approximate computation of the integral

$$I = \int_\Omega f(x)p(x)\,dx \tag{8.6}$$

where the nonnegative function $p(x)$ is the density function $\int_\Omega p(x)\,dx = 1$.

Let ξ be a random point with a probability density function $p(x)$. Introducing the random variable

$$\theta = f(\xi) \tag{8.7}$$

With mathematical expectation equal to the value of integral I

$$E(\theta) = \int_\Omega f(x)p(x)\,dx \tag{8.8}$$

Let the random points $\xi_1, \xi_2, \ldots, \xi_N$ be independent realizations of the random point ξ with the probability density function $p(x)$, then an approximate value of I is

$$\bar{\theta}_N = \frac{1}{N}\sum_{i=1}^{N} \theta_i \tag{8.9}$$

If $\bar{\xi}_N = \frac{1}{N}\sum_{i=1}^{N} \xi_i$ were absolutely convergent, then $\bar{\theta}_N$ would be convergent to I.

To combine MCS with the truss FEM, Figure 8.2 depicts the flowchart of the MC-FEM. It is clearly concluded in seven steps:

Step 1: Define the initial configuration of the graphene sheet, which includes corresponding parameters of bond length, height, and width of a hexagonal 2D lattice, and also thickness or diameter of the section.
Step 2: Fix material property parameters, which consist of Young's modulus, Poisson ratio, and physical density. The accuracy of the FEM for graphene sheets heavily depends on the exact parameters of material property.
Step 3: Apply the MCS method to randomly disperse vacancy defects in the graphene sheet in a certain percentage.

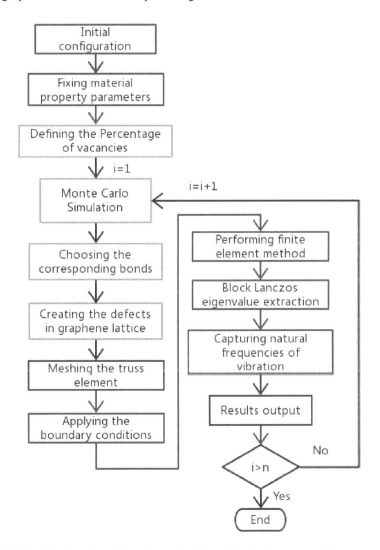

FIGURE 8.2   The flowchart of the Monte Carlo-based finite element method.

Step 4: By MCS, the corresponding number of chosen bonds is evident and can be used to create the vacancy defects in the graphene.

Step 5: Mesh the truss elements and apply the boundary conditions, which is a traditional step in the FEM.

Step 6: Perform the FEM and use Block Lanczos eigenvalue extraction for natural frequencies in each mode vibration of the graphene sheet.

Step 7: Capture the natural frequencies of graphene with vacancy defects in the vibration and output the results.

The loop continues until sufficient times of MCS are accomplished. Taking into consideration the effects of stochastically distributed vacancy, the loop is required to repeat, which provides reliable information.

8.4   GRAPHENE WITH STOCHASTIC DEFECTS

As mentioned above, randomly distributed vacancy defects were created in the lattice of monolayer graphene using a random number generator by the MCS. Each beam in the FEM of graphene sheets has a specific number. If a number is chosen in the MCS, a corresponding beam is removed from the entire hexagon structure of graphene sheets. The defects are quantified in terms of defect-density defined as:

$$Per = \frac{D_n}{A_n} \tag{8.10}$$

where D_n is the number of vacancy defects and A_n is the total number of beams in the FEM of the pristine graphene sheet.

For the parameters related to the material and geometrical properties, Young's modulus and Poisson's ratio in graphene sheets are settled as 1.2 TPa and 0.2, respectively; the length of the bond in the hexagon lattice is 0.27 nm, and the diameter in the cross section of the beam finite element is 0.032 nm. For the pristine graphene sheets, there are 6226 beams and 16,664 nodes created in the deterministic FEM. For the boundary condition, the six degrees of freedom for key points in the two longitudinal edges are all supposed to be zero. For each key point in the two transverse edges, they receive unit force in tension.

Figure 8.3 presents specific examples of graphene sheets with different percentages of vacancy defects. By the Monte Carlo method, the vacancy defects are randomly distributed in graphene sheets.

8.5   BUCKLING RESULTS AND DISCUSSION

8.5.1   PROBABILITY ANALYSIS

Given that vacancy defects are randomly distributed in graphene sheets, it is hard to evaluate or predict certain locations of vacancy defects throughout graphene sheets. When the number of vacancy defects, *Per*, is determined, one loop of MC-FEM is insufficient to represent the stochastic placement of vacancy defects. It is necessary to repeat the MCS and implement the FEM for sufficient times to include the main possible situations.

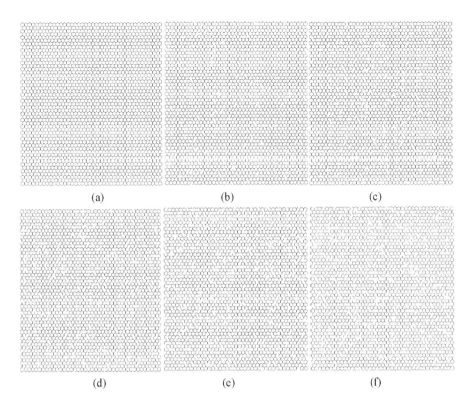

FIGURE 8.3 Specific examples of the graphene sheets with vacancy defects (a, b, c, d, e, and f have 2%, 4%, 6%, 8%, and 10% of vacancy defects, respectively).

Table 8.1 provides statistical results of the elastic buckling by repeating the MC-FEM. The average results of the critical buckling stress are more reliable and can be used to study the buckling behavior of vacancy-defected graphene sheets. Obviously, the mean of critical buckling is nonlinear with the increase of vacancy defects. When the *Per* exceeds 3%, the decrease of critical buckling stress is sudden. When the amount of vacancy defects is minor, the reduction is slow and insignificant.

TABLE 8.1
Statistical Results of the Elastic Buckling for the MC-FEM

Per (%)	Mean (THz)	Variance^0.5	Skewness	Kurtosis
1	5.6817	0.0294	−0.0009	0.0053
2	5.5534	0.0471	−0.0003	0.0056
3	5.0863	0.0996	−0.0004	0.0035
4	4.4207	0.2416	−0.0003	0.0037
5	3.1576	0.4646	−0.0003	0.0029

When *Per* is equal to 2%, compared with graphene without defects, the reduction rate of the critical buckling stress is 1.64%. The influence of vacancy defects on elastic buckling behavior is not obvious and can be ignored. However, when *Per* is 3%, 4%, and 5%, compared with graphene without defects, the reduction of the critical buckling stress is 9.91%, 21.7%, and 44.07%, respectively. Therefore, when the vacancy defects excess a certain density, the vacancy defects have a great influence on the elastic buckling behavior of graphene. The existence of vacancy defects damages the structural symmetry and integrity of graphene and has a deep effect on the buckling behavior.

As shown in Table 8.1, the critical buckling stress decreases in different degrees with the increase of vacancy defects. These results are considered by the vacancy positions in randomly dispersed graphene sheets based on the statistical results. Furthermore, the standard variance of the critical buckling stress increases with the increase of vacancy defects. When the vacancy defect percentage exceeds 3%, the augment of the standard variance of the critical buckling stress is also identically sharp.

Due to the randomly distributed placement of vacancy defects in graphene sheets, the critical buckling stress is deviated and fluctuated in certain interval ranges. Figure 8.4 manifests the probability density distribution of the critical buckling stress for graphene sheets with 1%, 2%, 3%, and 4% of vacancy defects, respectively. It is clear to find that for different amounts of vacancy defects, the histogram of the probability density distribution is approximated with the Gaussian or t location-scale distribution. The Gaussian or t location-scale distribution is more accurate for the description of the uncertainty and effects caused by stochastic vacancy defects.

In order to discuss the probability results more clearly, Figures 8.5 and 8.6 present the comparison of the probability density distribution and cumulative probability for graphene sheets with different vacancy defect percentages. It is obvious that when the vacancy defect amount is small, the influence of the randomly dispersed location of vacancy defects in the buckling behavior is weak. As shown in Figure 8.5, the probability density distribution of graphene sheets with 1% of vacancy defects is more concentrated than that of graphene sheets with 4%. Figure 8.6 also confirms this point. In a word, along with the increase of vacancy defects, the critical buckling stress of graphene sheets is distributed in a large interval with greater variance by the influence of stochastically dispersed vacancy defects.

8.5.2 COMPARISON AND DISCUSSION

Depending on the results of the MC-FEM, the probability analysis about the buckling behavior of vacancy-defected graphene sheets was successfully performed. Besides the probability results, the interval results of the critical buckling stress are also offered in this study. From the interval results, the extreme situations and the corresponding critical buckling stress are demonstrated in Figure 8.6. When the amount of vacancy defects is tiny, the maximum, the minimum, and the mean are clustered together or concentrated in a narrow range. This phenomenon well confirms the results of the probability density distribution and cumulative probability results in Figures 8.5 and 8.6.

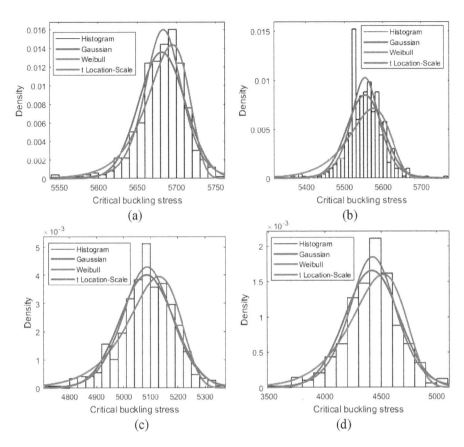

FIGURE 8.4 The probability density distribution for different percentage vacancy defects (a, b, c, and d are for 1%, 2%, 3%, and 4% of vacancy defects, respectively).

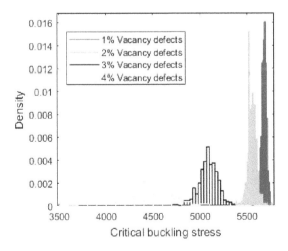

FIGURE 8.5 Comparison of the probability density for graphene sheets with different percentages of vacancy defects.

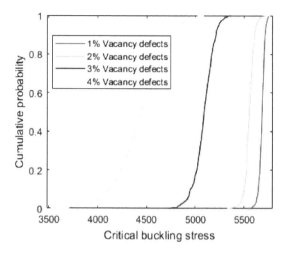

FIGURE 8.6   Comparison of the cumulative probability for graphene sheets with different percentages of vacancy defects.

In addition, the stiffness strengthening effects are observed in the curve of the maximum when the amount of vacancy defects is smaller than 2%, as marked in Figure 8.7. With the appearance of vacancy defects, the reduction of the critical buckling stress is not a unique impact. The augment of the critical buckling stress is possible to occur when *Per* is small. The curve of the maximum of critical buckling stress is not as monotonous as that of the minimum and the mean. It is clear that the curve of the maximum has two different stages. In the first stage, according to the augment of vacancy defects, the critical buckling stress becomes larger. However, in the second stage, along with the increase of vacancy defects, the critical buckling stress is sharply reduced. This strengthened circumstance is also measured and

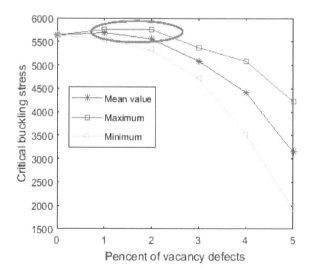

FIGURE 8.7   Interval results of MC-FEM.

tested in the physical experiments [31]. Good agreements are obtained in the results of the MC-FEM.

Furthermore, with the augment of the number of the vacancy defects, not only the critical buckling stress itself is effected, but also the interval between the maximum and minimum values of the MC-FEM is amplified. In Figure 8.8, the results of the presented method are compared with those of the molecular dynamics (MD) and finite element (FE) in the reported literature [32]. This demonstrates that the results in this study are generally smaller than the results of the MD and discrete FE, especially when the vacancy defect amount is small. The explanation for this deviation is from the perspective of vacancy types. The number of vacancy defects is counted by the number of vacancy beams in the hexagon lattice of graphene sheets in this study, while in the MD, the absence of atoms is recorded as vacancy defects. One atom of vacancy leads to three neighbor bonds lost reaction and connection. Besides, the vacancy defects in this study are dispersed in the entire graphene randomly. In the reported literature [32], the vacancy defects are periodically and regularly distributed. Therefore, it is reasonable to find that periodic atom vacancy defects have more distinct effects than the stochastic beam vacancy defects in the critical buckling stress of graphene sheets.

Parallel to this study, kinetic lattice MCS [33] is used to study the vacancy evolution in the entire graphene. By implementing ab initio energetics, the quantitative computation of the system kinetics, the morphology of defects, and their interactions are analyzed. The small aggregates and vacancy defects have evident influence in the nucleation stage of graphene. The vacancy defects in graphene are hampered by the relatively large barrier generated by the vacancy surrounding the strain field [34]. Therefore, the elastic buckling analysis of graphene is not only important to the mechanical properties, but also applicable to the kinetic vacancy evolution process.

The buckling of graphene is feasible and appropriate in the application of hydrogen storage and memcapacitor. The possibility of recruiting the buckled membrane as a plate of a capacitor with memory is validated by MD simulations and elastic mechanical calculations [35]. Besides, the storage and release of hydrogen are implemented

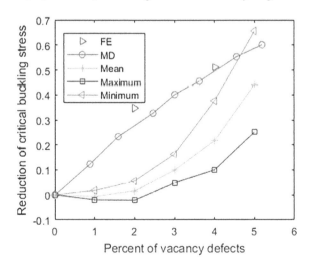

FIGURE 8.8 Comparison with results in the literature [32] (with 95% confidence interval).

in buckling graphene with convex and concave regions [36]. The buckling graphene is a revisable and environmental-friendly method of energy storage. Hydrogen chemisorption is energetically preferred in the convex regions of graphene, and the concave regions of graphene are more propitious to hydrogen release. By controlling the deformation of the elastic buckling in graphene, the dynamical and revisable process can be successfully conducted. The vacancy defects in graphene sheets can amplify the displacement and deformation of the graphene sheets in a specific location. The study of elastic buckling graphene with vacancy defects is a promising method to improve the efficiency of hydrogen storage.

8.5.3 DISPLACEMENT RESULTS OF GRAPHENE SHEETS

To demonstrate the buckling behavior of vacancy-defected graphene sheets, Figures 8.9 and 8.10 provide the vector sum results of the displacement in the first- and

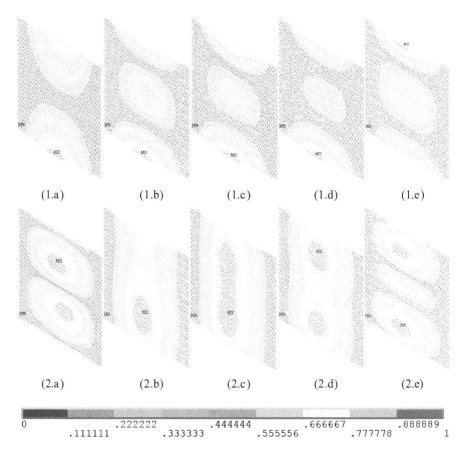

FIGURE 8.9   Vector sum of the displacement of graphene sheets (Figure 8.9.1 is for the first-order buckling mode; Figure 8.9.2 is for the fourth-order buckling mode; a, b, c, d, and e are for 0%, 1%, 2%, 3%, and 4% of vacancy defects, respectively).

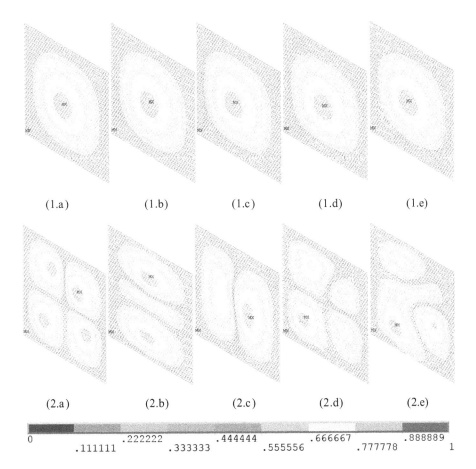

FIGURE 8.10 Vector sum of the displacement of graphene sheets (Figure 8.10.1 is for the first-order buckling mode; Figure 8.10.2 is for the fourth-order buckling mode; a, b, c, d, and e are for 0%, 1%, 2%, 3%, and 4% of vacancy defects, respectively).

fourth-order buckling modes. The boundary condition in Figure 8.9 is as mentioned above. The six degrees of freedom for key points in the two longitudinal edges are all supposed to be zero. For each key point in the two transverse edges, there is unit force in tension. In Figure 8.10, the situation of two longitudinal edges is the same as in Figure 8.10. Not only do the key points in the two transverse edges have unit force in tension, but also the rotations in each key points are limited.

Along with the increase of vacancy defects quantity, the vector sum of the displacement of the low-order buckling mode maintains geometrical symmetry. The change or deviation is not very evident, which can be found in Figures 8.9.1 Figure 8.10.1. However, for the high-order buckling mode, the results are totally contrary. Depending on the randomly distributed location of vacancy defects in graphene sheets, Figures 8.9.2.b–e are quite different from Figure 8.9.2.a. In Figure 8.10, the uncertainty in vacancy defects destroys the regular and symmetrical geometrical properties.

For the deformation of the elastic buckling, the results in this study have good agreements with quasi-static MD predictions. The buckling surfaces of highly cross-linked epoxy polymers under stress obey the paraboloid yield criterion at different temperatures [37]. Besides, by quasi-static simulations, the yield strength of the amorphous glassy polyethylene is computed in a hierarchical multiscale model with temperature and strain rate dependence [38]. The observation of transmission electron microscopy revealed that the buckling wavelengths are 3.6 ± 0.5 and $6.4 \pm 0.8\,\text{Å}$ in graphene, respectively [39]. There are only several (two or three) unit cells in the major buckling direction. Furthermore, the orientation of the lowest deformation energy is spontaneously chosen in the buckling process. The Euler buckling theory is appropriate in computing the deformation and displacement of graphene. Therefore, the MC-FEM is feasible to analyze the influence of randomly dispersed vacancy defects in the elastic buckling of graphene.

8.6   CONCLUSION

In this chapter, detailed studies on the elastic buckling of rectangular graphene sheets with different quantities of vacancy defects are carried out and performed by MC-FEM. The random dispersion of vacancy defects in graphene is taken into consideration, and the effects of the number of vacancy defects and stochastically distributed placements are discussed. From the proposed MC-FEM, the following conclusion can be drawn:

with the increase of vacancy defects in graphene, the critical buckling stress sharply decreases and the standard variance for the buckling stress evidently amplifies when the number of vacancy defects exceeds 3%. When the vacancy defect percentage is equal to 5%, the reduction of the critical buckling is as large as 44%. The vacancy defects profoundly influence the buckling behavior of the graphene lattice. With the increase of vacancy defects, the geometrical symmetry in the vector sum of displacement is obviously damaged.

The randomly distributed placement of vacancy causes fluctuation and deviation in the buckling behavior of the graphene vibration. The intervals between the maximum and minimum values are amplified along with the increase of vacancy defects. For the different amounts of vacancy defects, the probability density distributions of the critical buckling stress are all close to the Gaussian or t location-scale distribution.

Furthermore, the stiffness strengthening effect of vacancy defects in graphene is discussed. The possibility of mechanical property improvement in graphene by vacancy defects is confirmed.

REFERENCES

1. Kang, H., Kulkarni, A., Stankovich, S., et al. Restoring electrical conductivity of dielectrophoretically assembled graphite oxide sheets by thermal and chemical reduction techniques. *Carbon*, **2009**, 47(6), 1520–1525.
2. Bunch, J.S., Van Der Zande, A.M., Verbridge, S.S., et al. Electromechanical resonators from graphene sheets. *Science*, **2007**, 315(5811), 490–493.

3. Ferrari, A.C. Raman spectroscopy of graphene and graphite disorder, electron–phonon coupling, doping and nonadiabatic effects. *Solid State Communications*, **2007**, 143(1–2), 47–57.
4. Sakhaee-Pour, A., Ahmadian, M.T., Naghdabadi, R. Vibrational analysis of single-layered graphene sheets. *Nano-Technology*, **2008**, 19(8), 085702.
5. Mielke, S.L., Troya, D, Zhang, S. et al. The role of vacancy defects and holes in the fracture of carbon nanotubes. *Chemical Physics Letters* **2004**, 390(4), 413–420.
6. Tserpes, K.I., Vatistas, I. Buckling analysis of pristine and defected graphene. *Mechanics Research Communications*, **2015**, 64, 50–56.
7. Kim, J., Reddy, J.N. Analytical solutions for bending, vibration, and buckling of FGM plates using a couple stress-based third-order theory. *Composite Structures*, **2013**, 103(9), 86–98.
8. Togun, N., Bağdatli, S.M. Size dependent nonlinear vibration of the tensioned nano-beam based on the modified couple stress theory. *Composites Part B Engineering*, **2016**, 97, 255–262.
9. Xu, L., Ma, T.B., Hu, Y.Z., et al. Molecular dynamics simulation of the interlayer sliding behavior in few-layer graphene. *Carbon*, **2012**, 50(3), 1025–1032.
10. Ohta, Y., Okamoto, Y., Irle, S., et al. Rapid growth of a single-walled carbon nanotube on an iron cluster: Density-functional tight-binding molecular dynamics simulations. *ACS Nano*, **2008**, 2(7), 1437.
11. Gillen, R., Robertson, J. Density functional theory screened-exchange approach for investigating electronical properties of graphene-related materials. *Physical Review B*, **2010**, 82(12), 1303–1307.
12. Anjomshoa, A., Shahidi, A.R., Hassani, B., Jomehzadeh, E. Finite element buckling analysis of multi-layered graphene sheets on elastic substrate based on nonlocal elasticity theory. *Applied Mathematical Modelling*, **2014**, 38(24), 5934–5955.
13. Hosseini-Hashemi, S., Sharifpour, F., Ilkhani, M.R. On the free vibrations of size-dependent closed micro/nano spherical shell based on the modified couple stress theory. *International Journal of Mechanical Sciences*, **2016**, 115–116, 501–515.
14. Xu, W., Wang, L., Jiang, J. Strain gradient finite element analysis on the vibration of double-layered graphene sheets. *International Journal of Computational Methods*, **2016**, 13(03), 1650011.
15. Banhart, F., Kotakoski, J., Krasheninnikov, A.V. Structural defects in graphene. *ACS Nano*, **2010**, 5(1), 26–41.
16. Jing, N., Xue, Q., Ling, C., et al. Effect of defects on Young's modulus of graphene sheets a molecular dynamics simulation. *RSC Advances*, **2012**, 2(24), 9124–9129.
17. Xiao, J.R., Staniszewski, J., Gillespie Jr, J.W. Tensile behaviors of graphene sheets and carbon nanotubes with multiple Stone-Wales defects. *Materials Science and Engineering: A*, **2010**, 527(3), 715–723.
18. Ansari, R., Ajori, S., Motevalli, B. Mechanical properties of defective single-layered graphene sheets via molecular dynamics simulation. *Superlattices & Microstructures*, **2012**, 51(2), 274–289.
19. Namin, S., Asbaghian, F., Pilafkan, R. Vibration analysis of defective graphene sheets using nonlocal elasticity theory. *Physica E Low-dimensional Systems and Nanostructures*, **2017**, 93, 257–264.
20. Kim, B., Kim, T.W. Monte Carlo simulation for offshore transportation. *Ocean Engineering*, **2017**, 129, 177–190.
21. Olsen, B., Dufek, J. Stabilization effect of fission source in coupled Monte Carlo simulations. *Nuclear Engineering & Technology*, **2017**, 49(5), 1095–1099.
22. Leira, B.J., Næss, A., Næss, O.E.B. Reliability analysis of corroding pipelines by enhanced Monte Carlo simulation. *International Journal of Pressure Vessels & Piping*, **2016**, 144, 11–17.

23. Lee, H., Yoon, C., Cho, S., et al. The adaptation method in the Monte Carlo simulation for computed tomography. *Nuclear Engineering & Technology*, **2015**, 47(4), 472–478.

24. Chu, L., De Cursi, E.S., El Hami, A., Eid, M. Application of Latin hypercube sampling based kriging surrogate models in reliability assessment. *Science Journal of Applied Mathematics and Statistics*, **2015**, 3, 6.

25. Chu, L., De Cursi, E.S., El Hami, A., Eid, M. Reliability based optimization with meta-heuristic algorithms and Latin hypercube sampling based surrogate models. *Applied and Computational Mathematics*, **2015**, 4(6), 462–468.

26. Wernik, J.M., Meguid, S.A. Atomistic-based continuum modeling of the nonlinear behavior of carbon nanotubes. *Acta Mechanica*, **2010**, 212(1–2), 167–179.

27. Parvaneh, V., Shariati, M. Effect of defects and loading on prediction of Young's modulus of SWCNTs. *Acta Mechanica*, **2011**, 216(1–4), 281–289.

28. Brenner, D.W., et al. A second-generation reactive empirical bond order (REBO) potential energy expression for hydrocarbons. *Journal of Physics: Condensed Matter*, **2002**, 14(4), 783.

29. Tserpes, K.I. Strength of graphenes containing randomly dispersed vacancies. *Acta Mechanica*, **2012**, 223(4), 669–678.

30. Belytschko, T., et al. Atomistic simulations of nanotube fracture. *Physical Review B*, **2002**, 65(23), 235430.

31. Lópezpolín, G., Gómeznavarro, C., Parente, V., et al. Increasing the elastic modulus of graphene by controlled defect creation. *Nature Physics*, **2015**, 11(1), 26–31.

32. Sarvi, Z., Asgari, M., Shariyat, M., et al. Explicit expressions describing elastic properties and buckling load of BN nanosheets due to the effects of vacancy defects. *Superlattices & Microstructures*, **2015**, 88, 668–678.

33. Parisi, L., Giugno, R.D., Deretzis, I., et al. Kinetic Monte Carlo simulations of vacancy evolution in graphene. *Materials Science in Semiconductor Processing*, **2016**, 42, 179–182.

34. Trevethan, T., Latham, C.D., Heggie, M.I., et al. Vacancy diffusion and coalescence in graphene directed by defect strain fields. *Nanoscale*, **2014**, 6(5), 2978–2986.

35. Yamaletdinov, R.D., Ivakhnenko, O.V., Sedelnikova, O.V., et al. Snap-through transition of buckled graphene membranes for memcapacitor applications. *Scientific Report*, **2018**, 8(1), 3566.

36. Tozzini, V., Pellegrini, V. Reversible hydrogen storage by controlled buckling of graphene layers. *Journal of Physical Chemistry C*, **2011**, 115(51), 25523–25528.

37. Vubac, N, Bessa, M.A., Rabczuk, T., et al. A multiscale model for the quasi-static thermo-plastic behavior of highly cross-linked glassy polymers. *Macromolecules*, **2015**, 48(18), 6713–6723.

38. Vu-Bac, N, Areias, P.M.A, Rabczuk, T. A multiscale multisurface constitutive model for the thermo-plastic behavior of polyethylene. *Polymer*, **2016**, 105, 327–338.

39. Mao, Y., Wang, W.L., Wei, D., et al. Graphene structures at an extreme degree of buckling. *Acs Nano*, **2011**, 5(2), 1395–1400.

9 Impacts of Vacancy Defects in Resonant Vibration

9.1 INTRODUCTION

For the parameters corresponding to material properties of a defect-free graphene monolayer, Young's modulus and intrinsic strength are 1.0 TPa and 130 GPa, respectively [1]. A large deviation in simulations and experiments has been found, which is attributed to the presence and uncertainty of defects in the nanotube structure [2,3]. Various complicated defects exist in graphene sheets, and some defects in the atomic structure can even deteriorate the mechanical and electronic properties of graphene materials [4]. The research of vacancy defects in graphene sheets is very necessary, by which the uncertainties and fluctuation in mechanical property of graphene sheets can be reasonably explained and comprehensively understood.

Graphene, as a two-dimensional structure, was introduced in 2004 [5]. The main tough task in the analysis of the mechanical behavior of graphene is its small size [6–9]. Obviously, it is difficult to conduct physical experiments at the nanoscale. Analytical and numerical methods are available alternatives in nanomaterial research. These methods can be divided into two categories: atomistic-based methods and continuum mechanics theories. Atomistic-based methods include the molecular dynamics simulation (MD) [10], tight-binding molecular dynamics [11], and density functional theory [12]. Some size-dependent nonclassical continuum theories have been studied in the research field of graphene sheets, which includes the strain gradient theory [13], modified couple stress theory [14], and nonlocal elasticity theory [15]. However, atomistic-based methods are computationally expensive when the amount of atoms in the system is large. In continuum theories, the defects or uncertainties in the nanostructure are difficult to be taken into consideration. According to the overview of the existing literature, it can be found that the influence of the vacancy defects on the mechanical behavior of nanostructures has been studied in very few papers.

Attempts and struggles for a deeper understanding of the influence of defects in graphene sheets have been undertaken in this challenging field. For instance, the effect of vacancy defects on Young's modulus of graphene sheets was discussed by molecular dynamics [16]. The Stone–Wales defects in the tensile behavior of graphene sheets and carbon nanotubes were accomplished by the same method [17]. In addition, molecular dynamics simulations are applied in studying the mechanical property of defective single-layered graphene sheets [18]. The nonlocal elastic theory is also applied in vibration analysis of defective graphene sheets. However, the

DOI: 10.1201/9781003226628-11

effects of structural defects on the vibration behavior of graphene sheets have rarely been investigated [19].

The motivation of this study is to effectively analyze the vibration behavior of defected graphene sheets when the vacancy defects are randomly distributed. Since the physical experiments of vacancy-defected graphene sheets are difficult and expensive, numerical simulation methods are another promising supplement that needs to be developed. Besides, it is necessary to find an effective index to express and observe the effects of vacancy defects in the vibration behavior of graphene sheets.

In the free vibration, natural frequencies of structures are the essential parameters, which are directly corresponding with the geometrical and material characteristics. The change of the vacancy distribution and location can cause fluctuation in the natural frequencies of graphene sheets. The natural frequencies of vibration modes are very sensitive to geometrical defects. Therefore, natural frequency is a good indicator to discuss the effects of vacancy defects in graphene sheets.

Combining the Monte Carlo simulation with the finite element method to effectively simulate the randomly dispersed vacancy defects in graphene sheets is a perspective and feasible method. The Monte Carlo simulation as a sophisticated sampling method has been widely applied in the research of phase transition and magnetism of graphene sheets [20–23]. When the number of sampling is sufficient, it can reach satisfactory accuracy in numerical computations. It is common to take the results of the Monte Carlo simulation as the exact solution or comparison standard [24,25]. By combining the Monte Carlo method with the finite element model, an effective and feasible method is proposed. The difficulties in placement uncertainty, irregularity, and stochastic location are solved in the Monte Carlo simulation. The beam finite element model has well represented the specific hexagon microstructure of graphene sheets. The results computed by the proposed model are compared with those in the reported literature [26–34].

9.2   MATERIALS AND METHODS

The elastic geometrical properties of the beam elements have been derived using an energy relationship between molecular mechanics and continuum mechanics developed in Ref. [35]. According to the computation, the diameter d, Young's modulus E, and shear modulus G of the beam elements representing the C–C bonds are derived from

$$
\begin{cases}
d = 4\sqrt{\dfrac{k_\theta}{k_r}} \\[2ex]
E = \dfrac{k_r^2 L}{4\pi k_\theta} \\[2ex]
G = \dfrac{k_r^2 k_\tau L}{8\pi k_\theta^2}
\end{cases}
\tag{9.1}
$$

where k_r, k_θ, and k_τ are the bond sketching, bond bending, and torsional resistance force constants, respectively. In the finite element model of graphene sheets, the length of the C–C bond corresponds to the length of the beam in the planar-frame structure, and the wall thickness is related to the diameter of a circular solid cross section of the beam elements.

The beam elements applied in graphene sheets are based on the Timoshenko beam theory, which is a first-order shear deformation theory. In this theory, the transverse shear strain is constant through the cross section, which remains plane and undistorted after deformation.

The equation derived by Timoshenko that governs flexural vibrations of beams with constant cross section can be expressed as [36]

$$\frac{EI}{\rho A}\frac{\partial^4 \xi}{\partial z^4} - \frac{I}{A}\left(1+\frac{E}{\kappa G}\right)\frac{\partial^4 \xi}{\partial z^2 \partial t^2} + \frac{\partial^2 \xi}{\partial t^2} + \frac{\rho I}{\kappa G A}\frac{\partial^4 \xi}{\partial z^4} = 0 \tag{9.2}$$

where $\xi = \xi(z,t)$ is the transversal displacement along the x-axis at point z and time t, and E is Young's modulus, I is the inertia moment, G is the shear modulus, ρ is the mass density, and A is the cross-section area. In this theory, the Timoshenko shear coefficient κ is a free parameter.

Besides ξ, an angular variable θ is introduced. During flexural motion, cross sections are supposed to remain flat and perpendicular to the deflected neutral axis at any point of this axis. The angle θ between the z-axis and a vector orthogonal to the cross section is equal to the angle between the neutral axis tangent line and the z-axis. Note that θ equals the slope of the deflected neutral axis, that is,

$$\theta \approx \tan\theta = \frac{\partial \xi}{\partial z} \tag{9.3}$$

In a normal mode, $\xi(z,t)$ varies harmonically with time as

$$\xi(z,1) = [A\cos(wt) + B\sin(wt)\chi(z) = C\sin(wt+\varphi)\chi(z)] \tag{9.4}$$

where A, B, C, and φ are the corresponding constants to be determined. $w = 2\pi f$ is the angular frequency and $\chi(z)$ is a function that determines the normal mode amplitude. Substituting this form of ξ into Equation (9.2),

$$\frac{\partial^4 \chi}{\partial z^4} + \frac{\rho w^2}{M_r}\frac{\partial^2 \chi}{\partial z^2} + \frac{w^2 \rho^2}{\kappa GE}[w^2 - w_c^2]\chi = 0 \tag{9.5}$$

with $w_c = 2\pi f_c = \sqrt{\dfrac{\kappa GA}{\rho I}}$, where f_c is the critical frequency and $\dfrac{1}{M_r} = \left(\dfrac{1}{E}+\dfrac{1}{\kappa G}\right)$ is the reduced modulus.

It is well known that solutions of the above equation behave differently according to $w^2 - w_c^2$. The general solution can be written as

$$\begin{cases} \chi(z) = A_1 \sin(K_1 z) + B_1 \cos(K_1 z) + C_1 e^{K_2 z} + D_1 e^{-K_2 z} & w < w_c \\ \chi(z) = A_2 \sin(K_1 z) + B_2 \cos(K_1 z) + C_2 \sin(K_2 z) + D_2 \cos(K_2 z) & w > w_c \end{cases} \quad (9.6)$$

where

$$\begin{cases} K_1 = \sqrt{\dfrac{\rho w^2}{2M_r} + \sqrt{\left(\dfrac{\rho w^2}{2M_r}\right)^2 - \dfrac{\rho^2 w^2}{\kappa GE}(w^2 - w_c^2)}} \\ K_2 = \sqrt{S\left[\dfrac{\rho w^2}{2M_r} - \sqrt{\left(\dfrac{\rho w^2}{2M_r}\right)^2 - \dfrac{\rho^2 w^2}{\kappa GE}(w^2 - w_c^2)}\right]} \end{cases} \quad \text{with } S = \begin{cases} 1 & \text{if } w > w_c \\ -1 & \text{if } w < w_c \end{cases}$$

$$(9.7)$$

Usually, coefficients A_i, B_i, C_i, and D_i are different from zero, and the solutions of equations include functions depending on both K_1 and K_2. K_1 and K_2 are defined as positive square roots.

For free vibration analysis for Timoshenko beam, based on the principle of virtual work, the weak form of the equation can be written as [36]

$$\int_0^L EI \frac{\partial \theta}{\partial x} \delta\left(\frac{\partial \theta}{\partial x}\right) dx + \int_0^L \kappa GA\left(\frac{\partial \xi}{\partial x} - \theta\right)\delta\left(\frac{\partial \xi}{\partial x} - \theta\right) dx = \int_0^L \delta\xi \rho A \ddot{\xi} \, dx + \int_0^L \delta\theta \rho I \ddot{\theta} \, dx$$

$$(9.8)$$

As defined above, ξ is the transversal displacement in Timoshenko beam, where θ is the transversal rotation, while $\ddot{\xi}$ and $\ddot{\theta}$ are the transverse and rotary accelerations, respectively, L is the length of the beam, and δ denotes that the terms are virtual.

For free vibration,

$$\mathbf{Ku} - \mathbf{M\ddot{u}} = 0 \quad (9.9)$$

where $\mathbf{K}$ and $\mathbf{M}$ are the global stiffness and mass matrices which contain contributions from element stiffness and mass matrices.

A general solution can be expressed as

$$\mathbf{u} = \phi_k e^{iw_k} \quad (9.10)$$

Substituting into Equation (9.8) yields

$$[\mathbf{K} - w_k^2 \mathbf{M}]\phi_k = 0 \quad (9.11)$$

where ϕ_k is a set of displacement-type amplitude at the control points otherwise known as the model vector, and w_k is the natural frequency associated with the kth mode.

This is an eigenvalue problem and for nonzero solutions, the determinant of the equation must be zero.

$$\left| \mathbf{K} - w_k^2 \mathbf{M} \right| = 0 \tag{9.12}$$

The discretization of the governing equation can be done by a unidirectional linear finite element with two nodes. In each axis direction, there are two degrees of freedom at each node and six degrees of freedom for each node in total. The approximated solution in the displacement field, transversal displacement, and rotation can be written as [37]

$$\begin{cases} \xi = \displaystyle\sum_{i=1} \varphi_i \xi_i \\ \theta = \displaystyle\sum_{i=1} \varphi_i \theta_i \end{cases} \tag{9.13}$$

Consider the natural coordinate $\varsigma = [-1, 1]$ in the element domain,

$$\xi(\varsigma) = \varphi_1(\varsigma)\xi_1 + \varphi_2(\varsigma)\xi_2$$
$$\theta(\varsigma) = \varphi_1(\varsigma)\theta_1 + \varphi_2(\varsigma)\theta_2 \tag{9.14}$$

In the matrix form, it can be presented by

$$\begin{Bmatrix} \xi(\varsigma) \\ \theta(\varsigma) \end{Bmatrix} = [H]\{u\} \tag{9.15}$$

where $[H]$ is the matrix of shape functions and $\{u\}$ is the vector of nodal displacements.

$$[H] = \begin{bmatrix} \varphi_1 & 0 & \varphi_2 & 0 \\ 0 & \varphi_1 & 0 & \varphi_2 \end{bmatrix}, \quad \{u\} = \begin{Bmatrix} \xi_1 \\ \theta_1 \\ \xi_2 \\ \theta_2 \end{Bmatrix} \tag{9.16}$$

By substituting to the weak form of the equation, the stiffness and mass matrix in the element domain are written as

$$[K^e] = \int_{-1}^{1} [B]^T [D][B]|J| d\varsigma$$
$$[M^e] = \int_{-1}^{1} \rho I [H]^T [H]|J| d\varsigma + \int_{-1}^{1} \rho A [H]^T [H]|J| d\varsigma \tag{9.17}$$

where $[B]$ is the strain displacement matrix, $[D]$ is the constitutive matrix, and $|J|$ is the Jacobian determinant.

$$[B] = \begin{bmatrix} 0 & \dfrac{1}{|J|}\left(\dfrac{d\varphi_1}{d\varsigma}\right) & 0 & \dfrac{1}{|J|}\left(\dfrac{d\varphi_2}{d\varsigma}\right) \\[3ex] \dfrac{1}{|J|}\left(\dfrac{d\varphi_1}{d\varsigma}\right) & -\varphi_1 & \dfrac{1}{|J|}\left(\dfrac{d\varphi_2}{d\varsigma}\right) & -\varphi_2 \end{bmatrix} \qquad (9.18)$$

$$[D] = \begin{bmatrix} EI & 0 \\ 0 & \kappa GA \end{bmatrix}$$

For large symmetric eigenvalue problems, the Block Lanczos eigenvalue extraction method is popular and available [38]. A fast convergence rate can be reached by this method. In modal analysis, the Block Lanczos method is combined with Sturm sequence checks, and the number of eigenvalues requested is extracted by an automated shift strategy. In addition, the Sturm sequence check also makes sure that the requested number of eigenfrequencies beyond the user-supplied shift frequency is found without missing any modes [39].

9.3　VALIDATION OF THE MODEL

Using commercial software of ANSYS (Version 14.5, APDL, Cannonsburg, PA, USA), the typical graphene sheet lattice structure is created, which is in accordance with certain sizes mentioned in the literature. In the truss finite element model, there are 6226 trusses, 4212 points, and 16,664 nodes created. By performing the beam finite element model for graphene without vacancy defects, a good agreement between the present results and that of relevant references is achieved. Table 9.1 lists the corresponding parameters of the material property of graphene in literature, and in the proposed model, the results of natural frequencies for the first four modes of vibration.

From Table 9.1 and Figure 9.1, it is obvious that the present results of the beam finite element model have satisfied agreement with the corresponding results in the reported references. To be more exact, the natural frequencies of graphene in this study are close to the results of Reddy, Lu, Wei, Kudin, Liu, and Cadelano, slightly higher than the results of the molecular dynamics used by Khatibi, and lower than the results of Zhou and Gupta. The present beam finite element model is feasible to be applied in the subsequent study of vacancy-defected graphene sheets.

The vacancy defects are quantified according to the defect density, which is defined as

$$Per = \frac{D_n}{A_n} \qquad (9.19)$$

where D_n is the amount of vacancy defects and A_n is the total quantity of beams in graphene.

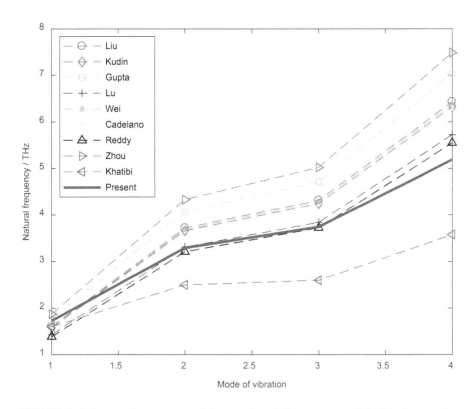

FIGURE 9.1   Comparison of natural frequencies of the present model with those in the literature.

TABLE 9.1
Comparing the Natural Frequencies Obtained from the Present Model with Those in References [26–34]

Reference	Method	E(TPa)	v	1/THz	2/THz	3/THz	4/THz
Fang et al. [26]	DFT	1.050	0.186	1.6081	3.7232	4.3172	6.4323
Kudin et al. [27]	DFT	1.029	0.149	1.5818	3.6623	4.2466	6.3271
Gupta et al. [28]	MD	1.272	0.147	1.7581	4.0706	4.7201	7.0325
Lu and Rui [29]	MD	0.725	0.398	1.4311	3.3135	3.8422	5.7246
Wei et al. [30]	DFT	1.039	0.169	1.5946	3.6921	4.2811	6.3786
Cadelano et al. [31]	TB	0.931	0.310	1.5649	3.6232	4.2012	6.2595
Reddy et al. [32]	MM	0.669	0.416	1.3869	3.2111	3.7234	5.5475
Zhou et al. [33]	MM	1.167	0.456	1.8716	4.3334	5.0248	7.4865
Sadeghzadeh et al. [34]	MD + FDD	1.050	0.170	1.6030	2.4970	2.5980	3.5770
Present	SFEM	1.200	0.200	1.7282	3.2925	3.7442	5.1892

*E and v are Young's modulus and Poisson's ratio, respectively.

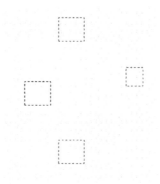

FIGURE 9.2   Randomly distributed vacancy in a graphene sheet by Monte Carlo simulation.

The relative error is used to calculate the difference of the mean of the natural frequencies with that of graphene without defects. Error can be written as

$$\text{Error} = \text{abs}(P_f - M_f) \tag{9.20}$$

where M_f is the mean of natural frequencies of vacancy-defected graphene, and P_f is the natural frequency of pristine graphene.

Figure 9.2 depicts a possible example for the Monte Carlo-based stochastic distributed placement of vacancy in graphene sheets.

9.4   RESULTS AND DISCUSSION

In the following part, the explicit discussion about the amount of vacancy defects and parameters corresponding to geometrical and material characteristics of graphene sheets is sequentially presented. The results are computed by the Monte Carlo-based finite element method and demonstrated by the probability theory and mathematical statistics.

9.4.1   AMOUNT OF VACANCY DEFECTS

The amount of vacancy defects accounted for in this study is by the absence of beam in the stochastic placement. Table 9.2 provides the statistical results of the first four-order natural frequencies of vibration modes for graphene sheets with different numbers of vacancy defects. By taking the uncertainty of the randomly distributed location of vacancy defects into consideration, the probability and statistical results of natural frequencies are more reliable and suitable for the study of vibration behavior of graphene sheets. It is clear that, with the increase of vacancy defects, the mean of the first four-order natural frequencies is diminished. When *Per* is 0.5% and 1%, the changes of natural frequencies caused by randomly distributed vacancy defects are not evident. However, the reduction of natural frequencies is abrupt when the percentage of vacancy defects exceeds 1%.

The natural frequencies of the beam finite element model as derived are corresponding to the stiffness and mass matrices. The fluctuation in natural frequencies in different vibration modes is a comprehensive effect of change in Young's modulus and mass reduction by vacancy defects. According to the relative reference [40], the elastic modulus of graphene can increase by well-controlled defect creation. Young's modulus reaches the peak when the percentage of vacancy defects of carbon atoms is 0.2%. It is reasonable to find that when *Per* equals to 0.5% and 1%, the changes of natural frequencies (Error) in Table 9.2 and Figure 9.3c are not evident.

When *Per* is 3%, the reduction of the first four-order natural frequencies compared with that of pristine graphene are 7.1%, 8.7%, 6.04%, and 8.7%, respectively. Besides, the reduction reaches 32.1%, 46.7%, 41.3%, and 53.1% for the first four-order vibration mode when *Per* equals to 5%. As shown in Table 9.2 and Figure 9.3, with the increase of vacancy defects, the first four natural frequencies all have different degrees of reduction. Besides, the standard variance of natural frequencies becomes larger along with the increase of *Per*. Furthermore, in the first four vibration modes, the standard variances increase sharply when *Per* is above 1%. The vacancy defects greatly affect the vibration behavior of graphene sheets. The existence of vacancy defects changes the structure of graphene lattice and has a profound impact on natural frequencies.

As mentioned above, natural frequencies are the ratios of stiffness and mass matrices. The appearance of vacancy defects influences both the stiffness and mass matrices. The mass of graphene sheets decreases according to the increase in the amount

TABLE 9.2
Statistical Results of Natural Frequencies for the Monte Carlo-Based Finite Element Model

Per (%)	Mode	Mean (THz)	Variance^0.5	Skewness	Kurtosis
0.5	1	1.721	0.002	−0.852	5.179
	2	3.279	0.004	−0.646	3.751
	3	3.729	0.006	−0.536	3.935
	4	5.168	0.005	−0.620	3.625
1	1	1.714	0.003	−0.446	3.144
	2	3.265	0.006	−0.559	3.558
	3	3.713	0.008	−0.495	3.537
	4	5.146	0.007	−0.308	2.757
3	1	1.602	0.292	−5.313	29.246
	2	3.007	0.547	−5.312	29.240
	3	3.512	0.639	−5.312	29.242
	4	4.730	0.861	−5.312	29.243
5	1	1.172	0.584	−1.488	3.247
	2	1.752	0.871	−1.491	3.270
	3	2.196	1.090	−1.501	3.300
	4	2.429	1.204	−1.509	3.316

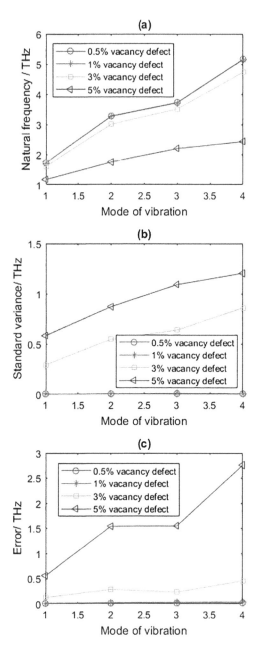

FIGURE 9.3 Results of defected graphene sheets with different amounts of vacancy defects (a–c for mean, standard variance, and error of natural frequency, respectively).

of vacancy defects. The influence of vacancy defects in Young's modulus of graphene sheets is not monotonous and nonlinear [31,41]. When the number of vacancy defects is small, by effectively controlling the distribution of vacancy defects, the augmentation of the amount of vacancy defects in graphene sheets can increase Young's modulus [40]. However, when the amount of vacancy defects reaches a certain value, Young's modulus dramatically reduces. The augmentation of vacancy defects leads to a cut down not only in Young's modulus, tensile strength, and failure strain, but also in fracture propagation and crack formation [42]. Vibration analysis is an appropriate and feasible way to observe the influence of the amount of vacancy defects in graphene sheets.

9.4.2 GEOMETRICAL PARAMETERS

On the basis of molecular mechanics by the energy approach, the tube diameter and length strongly affect the material properties, such as Young's modulus, shear modulus, bend-angle, and strain energy of single- and multi-walled nanotubes [43]. The discussion about geometrical parameters, namely the bond length L, the diameter of cross section D, and the number of hexagons in the y-axis N, is important and necessary in the vibration analysis of graphene sheets.

The results are based on the statistical results by consideration of the random dispersion of vacancy location in the graphene sheets. In Figure 9.4, with the amplification of L, the mean of the first four-order natural frequencies is distinctly cut down. The standard variance and error in Figures 9.4b and c also follow this rule. When the amount of the vacancy defects is 1% and L is 0.15, with the augmentation of the length of each bond in the hexagon of graphene sheets, the increase of the first four-order natural frequencies compared with that of graphene sheets in Table 9.2 are 223.98%, 224.01%, 224.00%, and 223.90%, respectively. When L is equal to 0.2, the increases are 82.26%, 82.27%, 82.25%, and 82.22%, respectively. When L is 0.32, the reductions of the first four-order natural frequencies of vibration modes are 28.82%, 28.79%, 28.82%, and 28.82%, respectively. Therefore, the length of bond in the hexagon of graphene sheets is a crucial factor, which shows a good agreement of size effects in physical experiments and numerical simulation.

According to the results above, the geometrical parameter L has a negative effect on the first four-order natural *frequencies*. The larger the L, the smaller the natural frequencies. Different from the effects of L in the vibration behavior of vacancy-defected graphene sheets, with the increase of D, the first four-order natural frequencies are amplified. Figure 9.5 well proves this point. The standard variance in Figure 9.5b also follows this rule, but the results of errors in Figure 9.5c are different. Compared with the results in Table 9.2 when *Per* equals to 1% and D is 0.02, the reduction of the first four-order natural frequencies of vibration modes are 37.51%, 37.49%, 37.49%, and 37.50%, respectively. When D is equal to 0.026, the reduction of the first four-order natural frequencies of vibration modes compared with that of graphene in Table 9.2 is 18.73%, 18.74%, 18.74%, and 18.75%, respectively. When L is 0.04, the increases of the first four-order natural frequencies of vibration modes are 24.97%, 25.02%, 24.97%, and 24.99%, respectively. Both the cross section and inertia moment are related to the diameter of cross section D, which has more complicated effects in natural frequencies of vibration modes. Even though the change of extent in natural frequencies is not as large as the effect of the bond length L, more emphasis should be put on the diameter of cross section D.

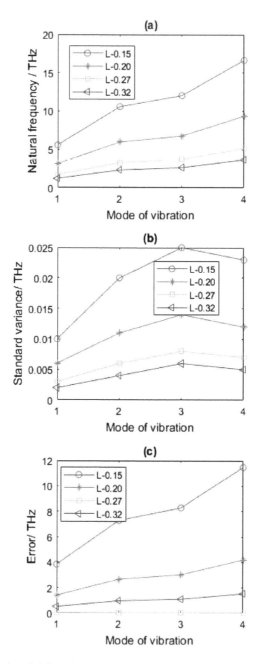

FIGURE 9.4 Results of defected graphene sheets with different L (a–c for mean, standard variance, and error of natural frequency, respectively).

There is another geometrical parameter, N, the number of hexagons in y-axes, which is not only related to the width of graphene sheets, but also corresponding with the amount of vacancy defects if Per is determined as a certain value. Compared

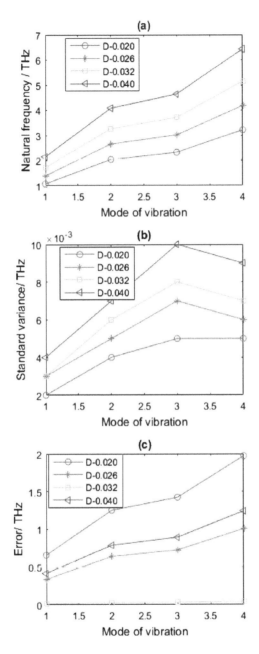

FIGURE 9.5   Statistical results of defected graphene sheets with different D (a–c for mean, standard variance, and error of natural frequency, respectively).

with the results in Table 9.2 and Figure 9.6, when *Per* equals to 1% and N is 30, the natural frequencies of the first four orders of vibration mode are 47.90%, 20.83%, 66.55%, and 22.85% larger, respectively. When N is equal to 35, the increases of the natural frequencies of the first four vibration modes compared with that of graphene

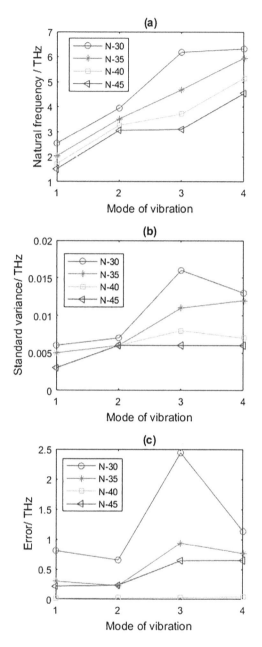

FIGURE 9.6 Statistical results of defected graphene sheets with different N (a–c for mean, standard variance, and error of natural frequency, respectively).

in Table 9.2 are 18.32%, 7.81%, 26.04%, and 15.68%, respectively. When N is 45, the reductions of the natural frequencies of the first four-order vibration modes are 11.84%, 6.34%, 16.43%, and 11.76%, respectively.

In short, different from the effects of bond length L and the diameter of cross section D, the influence of N in natural frequencies of vacancy-defected graphene sheets is more intricate and disproportionate, as in Figure 9.6b and c. In general, more emphasis is put on the bond length and the diameter of the cross section, and the total width and length of graphene sheets are controlled and limited by experimental operation and chemical methods. However, the number of hexagons in the axis directions is another sensitive factor related to the natural frequencies of vacancy-defected graphene sheets.

9.4.3 Material Parameters

A large deviation between simulations and experiments in material properties has been found because of the presence of defects in the nanotube structure [2,44]. The mechanical properties of defect-free carbon nanomaterials have not been exactly measured yet. For discussion, material parameters (Young's modulus and Poisson's ratio of graphene sheets) are supposed to be specific values.

From Figure 9.7, it is distinct to find that, with the increase of E, namely Young's modulus, the mean of the first four natural frequencies evidently increased as well. The standard variance and error in Figure 9.7b also follow this rule. However, the results in Figure 9.6c are more complicated. When the amount of the vacancy defects is fixed to 1%, and E is 800 GPa, the reductions of the natural frequencies of the first four vibration modes compared with that of graphene sheets in Table 9.2 are 18.38%, 18.35%, 18.34%, and 18.36%, respectively. When E is equal to 1500 GPa, the increase in the natural frequencies of the first four vibration modes compared with that of pristine graphene is 11.79%, 11.82%, 11.80%, and 11.80%, respectively. When E is 2000 GPa, the increments of the natural frequencies of the first four vibration modes are 29.11%, 29.13%, 29.11%, and 29.09%, respectively.

With the augment of Young's modulus, the values in the stiffness matrix of vacancy-defected graphene sheets are amplified, while the stiffness matrix acts as the numerator, which is reasonable to find the enlargement of natural frequencies in vibration modes. Furthermore, as mentioned in the above results, the first four natural frequencies of graphene sheets all have a close magnitude of change. The effects of Young's modulus in natural frequencies are convergent to the specific percentage in the different vibration modes.

Poisson's ratio as one of the parameters corresponding to the material properties is discussed in Figure 9.8. The change of the Poisson ratio in vacancy-defected graphene sheets does not play a significant role in the natural frequencies of graphene sheets. As shown in Figure 9.8a, with the increase of V, the results of natural frequencies are very approaching. In other words, the Poisson ratio can be simplified as a neglected factor in the analysis of natural frequencies for vacancy-defected graphene sheets.

9.4.4 Graphene Sheets with Vacancy Defects

To observe the vibration behavior of graphene with vacancy defects, Figures 9.9 and 9.10 present the displacement vector sum and rotation vector sum for graphene with

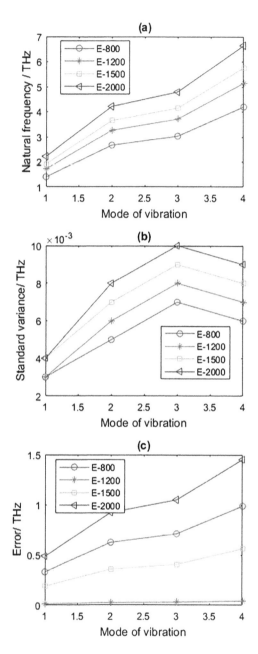

FIGURE 9.7 Statistical results of defected graphene sheets with different Young's modulus (a–c for mean, standard variance, and error of natural frequency, respectively). E is Young's modulus of graphene sheets.

5% vacancy defects, respectively. Along with the augment of vacancy defects in the graphene sheet, the natural frequencies of vibration mode are evidently cut down. As demonstrated in Figures 9.9 and 9.10, the regular and symmetrical property is also

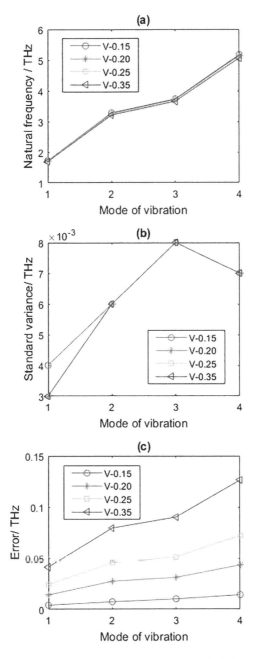

FIGURE 9.8 Statistical results of defected graphene sheets with different Poisson ratios (a–c for mean, standard variance, and error of natural frequency, respectively). *V* is the Poisson ratio.

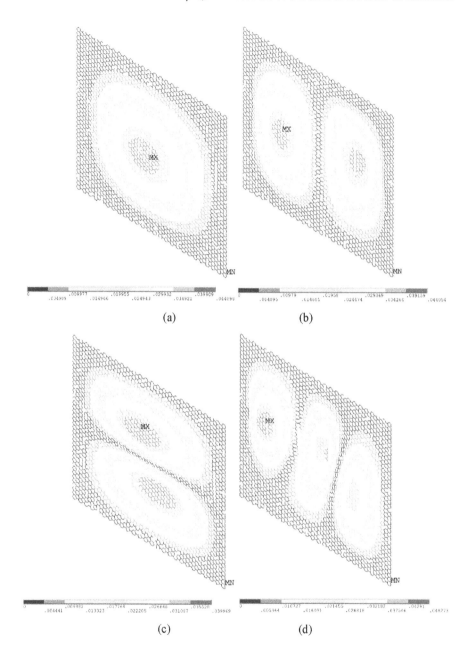

FIGURE 9.9 Displacement vector sum for graphene with 5% vacancy defects (a–d show the first order, second order, third order, and fourth order of vibration mode, respectively).

destroyed because of the randomly distributed vacancy defects. Therefore, besides the change of the first four-order natural frequencies in the vibration behavior of graphene sheets, the vacancy defects can also deeply influence the displacement and rotation vector of graphene sheets.

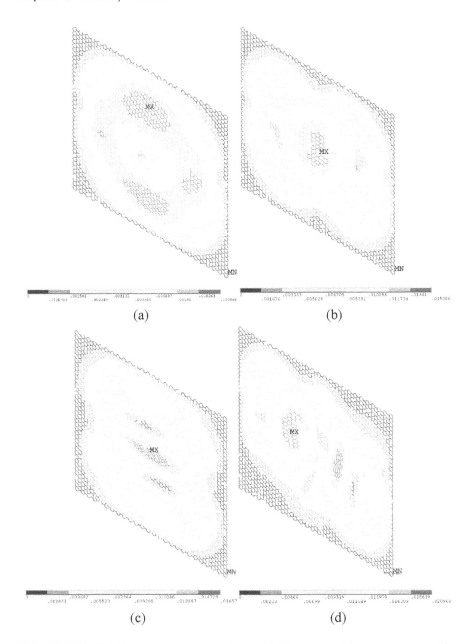

FIGURE 9.10   Rotation vector sum for graphene with 5% vacancy defects (a–d represent the first, second, third, and fourth orders of vibration mode, respectively).

In addition, besides the effects of vacancy defects, impurities in graphene sheets are another unavoidable existence and need more attention [45]. Natural frequencies of vacancy-defected graphene sheets are a sensitive and appropriate index to comprehensively discuss and study the influence of vacancy defects in stiffness and mass matrices of entire graphene sheets. In addition to mechanical properties, thermal and electronic transport properties are promising in perspective for vacancy-defected graphene sheets [46].

9.5 CONCLUSIONS

In this chapter, an effective Monte Carlo-based finite element method to analyze the vibration behavior of vacancy-defected graphene sheets is proposed. By combining the Monte Carlo method with the beam FE model, the stochastic dispersed placement of vacancies in graphene is successfully propagated and performed. The results show that

1. The natural frequencies of vibration decrease dramatically with the increase of the vacancy defects when *Per* is larger than 1%.
2. The regular and symmetrical properties of vibration behavior in displacement and rotation are evidently destroyed when *Per* reaches 5%.
3. Furthermore, with the augment of the vacancy defects quantity, the variances caused by the random dispersion location of vacancy defects become large.
4. From the discussion about geometrical and material parameters, it can be concluded that the increment of length of each bond in hexagons of graphene sheets distinctly reduces the first four-order natural frequencies.
5. For vacancy-defected graphene sheets, the amount of hexagons in the long side of the graphene sheet has a similar effect on natural frequencies with the length of each bond in hexagon, but it is more complicated.
6. With the increase of thickness of the graphene sheets and Young's modulus, the natural frequencies of vacancy-defected graphene become smaller.

REFERENCES

1. Lee, C., Wei, X., Kysar, J.W., James, H. Measurement of the elastic properties and intrinsic strength of monolayer graphene. *Science*, **2008**, 321, 385–388.
2. Mielke, S.L., Troya, D., Zhang, S., Li, L.-L., Xiao, S., Car, R., Ruoff, R.S., Schatz, G.C., Belytschko, T. The role of vacancy defects and holes in the fracture of carbon nanotubes. *Chemical Physics Letters*, **2004**, 390, 413–420.
3. Tserpes, K.I., Vatistas, I. Buckling analysis of pristine and defected graphene. *Mechanics Research Communications*, **2015**, 64, 50–56.
4. Banhart, F., Jani, K., Krasheninnikov, A.V. Structural defects in graphene. *ACS Nano*, **2011**, 5, 26–41.
5. Novoselov, K.S., Geim, A.K., Morozov, S.V., Jiang, D., Zhang, Y., Dubonos, S.V., Grigorieva, I.V., Firsov, A.A. Electric field effect in atomically thin carbon films. *Science*, **2004**, 306, 666–669.
6. Thai, H.-T., Vo, T.P. A nonlocal sinusoidal shear deformation beam theory with application to bending, buckling, and vibration of nanobeams. *International Journal of Engineering Science*, **2012**, 54, 58–66.
7. Kim, J., Reddy, J.N. Analytical solutions for bending, vibration, and buckling of FGM plates using a couple stress-based third-order theory. *Composite Structure*, **2013**, 103, 86–98.

8. Togun, N., Bağdatli, S.M. Size dependent nonlinear vibration of the tensioned nano-beam based on the modified couple stress theory. *Composites Part B: Engineering*, **2016**, 97, 255–262.

9. Kambali, P.N., Nikhil, V.S., Pandey, A.K. Surface and nonlocal effects on response of linear and nonlinear NEMS devices. *Applied Mathematical Modelling*, **2016**, 43, 252–267.

10. Nazemnezhad, R., Zare, M., Hosseini-Hashemi, S., Shokrollahi, H. Molecular dynamics simulation for interlayer interactions of graphene nanoribbons with multiple layers. *Superlattices and Microstructures*, **2016**, 98, 228–234.

11. Maslov, M.M., Katin, K.K. High kinetic stability of hypercubane: Tight-binding molecular dynamics study. *Chemical Physics Letters*, **2016**, 644, 280–283.

12. Lee, H.W., Moon, H.S., Hur, J., Kim, I.T., Park, M.S., Yun, J.M., Kim, K.H., Lee, S.G. Mechanism of sodium adsorption on N-doped graphene nanoribbons for sodium ion battery applications: A density functional theory approach. *Carbon*, **2017**, 119, 492–501.

13. Kundalwal, S.I., Meguid, S.A., Weng, G.J. Strain gradient polarization in graphene. *Carbon*, **2017**, 117, 462–472.

14. Hosseini-Hashemi, S., Sharifpour, F., Ilkhani, M.R. On the free vibrations of size-dependent closed micro/nano spherical shell based on the modified couple stress theory. *International Journal of Mechanical Sciences*, **2016**, 115–116, 501–515.

15. Anjomshoa, A., Shahidi, A.R., Hassani, B., Jomehzadeh, E. Finite element buckling analysis of multi-layered graphene sheets on elastic substrate based on nonlocal elasticity theory. *Applied Mathematical Modelling*, **2014**, 38, 5934–5955.

16. Jing, N., Xue, Q., Ling, C., Shan, M., Zhang, T., Zhou, X., Jiao, Z. Effect of defects on Young's modulus of graphene sheets: A molecular dynamics simulation. *RSC Advances*, **2012**, 2, 9124–9129.

17. Xiao, J.R., Staniszewski, J., Gillespie, J.W., Jr. Tensile behaviors of graphene sheets and carbon nanotubes with multiple Stone–Wales defects. *Materials Science and Engineering A*, **2010**, 527, 715–723.

18. Ansari, R., Ajori, S., Motevalli, B. Mechanical properties of defective single-layered graphene sheets via molecular dynamics simulation. *Superlattices and Microstructures*, **2012**, 51, 274–289.

19. Asbaghian Namin, S.F., Pilafkan, R. Vibration analysis of defective graphene sheets using nonlocal elasticity theory. *Physics E*, **2017**, 93, 257–264.

20. Ulybyshev, M.V., Buividovich, P.V., Katsnelson, M.I., Polikarpov, M.I. Monte Carlo study of the semimetal-insulator phase transition in monolayer graphene with a realistic interelectron interaction potential. *Physical Review Letters*, **2013**, 111, 056801.

21. Armour, W., Simon, H., Strouthos, C. Monte Carlo simulation of the semimetal-insulator phase transition in monolayer graphene. *Physical Review B*, **2010**, 81, 125105.

22. Feldner, H., Meng, Z.Y., Honecker, A., Cabra, D., Wessel, S., Assaad, F.F. Magnetism of finite graphene samples: Mean-field theory compared with exact diagonalization and quantum Monte Carlo simulations. *Physical Review B*, **2010**, 81, 115416.

23. Whitesides, R., Frenklach, M. Detailed kinetic Monte Carlo simulations of graphene-edge growth. *Journal of Physical Chemistry A*, **2009**, 114, 689–703.

24. Liu, C., Cursi, E.S.D., Hami, A.E., Eid, M. Application of Latin hypercube sampling based kriging surrogate models in reliability assessment. *Science Journal of Applied Mathematics and Statistics*, **2015**, 3, 263–274.

25. Liu, C., Cursi, E.S.D., Hami, A.E., Eid, M. Reliability based optimization with meta-heuristic algorithms and Latin hypercube sampling based surrogate models. *Applied and Computational Mathematics*, **2015**, 4, 462–468.

26. Fang, L., Pingbing, M., Li, J. Ab initio calculation of ideal strength and phonon instability of graphene under tension. *Physical Review B*, **2007**, 76, 064120.

27. Kudin, K.N., Scuseria, G.E., Yakobson, B.I. C_2F, BN, and C nanoshell elasticity from ab initio computations. *Physical Review B*, **2001**, 64, 235406.
28. Gupta, S., Dharamvir, K., Jindal, V.K. Elastic moduli of single-walled carbon nanotubes and their ropes. *Physical Review B*, **2005**, 72, 165428.
29. Lu, Q., Rui, H. Nonlinear mechanics of single-atomic-layer graphene sheets. *International Journal of Applied Mechanics*, **2009**, 1, 443–467.
30. Wei, X., Fragneaud, B., Marianetti, C.A., Kysar, J.M. Nonlinear elastic behavior of graphene: Ab initio calculations to continuum description. *Physical Review B*, **2009**, 80, 205407.
31. Cadelano, E., Palla, P.L., Giordano, S., Colombo, L. Nonlinear elasticity of monolayer graphene. *Physical Review Letters.* **2009**, 102, 235502.
32. Reddy, C.D., Rajendran, S., Liew, K.M. Equilibrium configuration and continuum elastic properties of finite sized graphene. *Nanotechnology*, **2006**, 17, 864.
33. Zhou, L., Wang, Y., Cao, G. Elastic properties of monolayer graphene with different chiralities. *Journal of Physics: Condensed Matter*, **2013**, 25, 125302.
34. Sadeghzadeh, S., Khatibi, M.M. Modal identification of single layer graphene nanosheets from ambient responses using frequency domain decomposition. *European Journal of Mechanics A/Solids*, **2017**, 65, 70–78.
35. Tserpes, K.I., Papanikos, P. Finite element modeling of single-walled carbon nanotubes. *Composites B Engineering*, **2005**, 36, 468–477.
36. Sang, J.L., Park, K.S. Vibrations of Timoshenko beams with isogeometric approach. *Applied Mathematical Modelling*, **2013**, 37, 9174–9190.
37. Shang, H.Y. Enriched finite element methods for Timoshenko beam free vibration analysis. *Applied Mathematical Modelling*, **2016**, 40, 7012–7033.
38. Rajakumar, C. Lanczos algorithm for the quadratic eigenvalue problem in engineering applications. *Computer Methods in Applied Mechanics and Engineering.* **1993**, 105, 1–22.
39. Cullum, J., Willoughby, R.A. A survey of Lanczos procedures for very large real 'symmetric' eigenvalue problems. *Journal of Computational and Applied Mathematics.* **1985**, 12–13, 37–60.
40. López-Polín, G., Gómez-Navarro, C., Parente, V., Guinea, F., Katsnelson, M.I., Perez-Murano, F., Gomez-Herrero, J. Increasing the elastic modulus of graphene by controlled defect creation. *Nature Physics*, **2014**, 11, 26–31.
41. Dettori, R., Cadelano, E., Colombo, L. Elastic fields and moduli in defected graphene. *Journal of Physics: Condensed Matter*, **2012**, 24, 104020.
42. Tserpes, K.I. Strength of graphenes containing randomly dispersed vacancies. *Acta Mechanica*, **2012**, 223, 669–678.
43. Shen, L., Jackie, L. Transversely isotropic elastic properties of multiwalled carbon nanotubes. *Physical Review B* **2005**, 71, 035412.
44. Tserpes, K.I., Papanikos, P. The effect of Stone–Wales defect on the tensile behavior and fracture of single-walled carbon nanotubes. *Composite Structure*, **2007**, 74, 581–589.
45. Pellegrino, F.M.D., Angilella, G.G.N., Pucci, R. Effect of impurities in high-symmetry lattice positions on the local density of states and conductivity of graphene. *Physical Review B*, **2009**, 80, 094203.
46. Parisi, L., Angilella, G.G.N., Deretzis, I., Renato, P., Magna, A.L. Role of H distribution on coherent quantum transport of electrons in hydrogenated graphene. *Condensed Matter*, 2017, 2, 37.

10 Uncertainty Quantification in Nanomaterials

10.1 INTRODUCTION

With a two-dimensional (2D) honeycomb lattice, graphene can be wrapped up into zero-dimensional (0D) fullerenes, rolled into one-dimensional (1D) nanotubes, or stacked into three-dimensional (3D) graphite [1]. Graphene has extraordinary properties for potential applications. The electrical conduction of graphene is proved to be possible and discovered in the mechanical exfoliation samples [2–4]. Besides, the measurements of thermal properties of graphene [5–7] also confirm the distinguished thermal conductivity of graphene and therefore ignite strong interests of researchers. The unexpected mechanical properties of graphene are experimentally verified through nanoindentation by atomic force microscopy (AFM) [8].

It is recognized in academia that graphene has astonishing strength and stiffness when compared with traditional materials. However, the precise values of the parameters corresponding to the material properties of graphene are difficult to be determined. On the aspects of experimental measurements, the in-plane Young's modulus of bulk graphite [9] is in the range of 1.02 ± 0.03 TPa. In the tensile test [10], a broad range of stiffness values (0.27–1.47 TPa) were obtained, with breaking strengths ranging from 3.6 to 63 GPa. In addition, the defects in the graphene contribute to the deviation in the bending rigidity in the test results of suspended monolayer graphene membranes [11]. Furthermore, even when the same test method is used in the measurements, the results obtained by researchers are different. For example, Young's modulus is extracted as 0.5 TPa [12] in the measurement of the bending stiffness of graphene sheets by AFM nanoindentation. In another study, Young's modulus is equal to 1.0 ± 0.1 TPa and the corresponding intrinsic stress is 130 ± 10 GPa at a strain of 0.25 [8].

The fluctuation and deviation in the parameters related to the material properties of graphene are owing to the inevitable uncertainties. On the one hand, the nanoscale size in graphene makes the exact measurement challenging, and the relative errors caused by equipment or other stochastic factors are amplified and significantly affect the final results. On the other hand, the mathematical expressions and related knowledge at the macroscale are limited and not appropriate to describe the physical relationships in a microscope. The definition and evaluation of certain parameters at the microscale are required to be developed with accurate recognitions.

Different from epistemic uncertainties, the aleatory uncertainties in graphene are also unavoidable in real situations. First, the randomly distributed defects contribute to the variation of material properties in graphene [13]. Second, the grain boundaries in macroscopic samples are hard to control. The mechanical parameters are sensitive to the size and shape of grain boundaries of graphene sheets. Third, uncertainties in the sample geometry, stress concentration at clamping points, and unknown load distributions are all the uncertain problems confronted by researchers. Therefore, it is necessary to develop a feasible model which takes the uncertainties into consideration for the analysis of graphene.

The development of sophisticated models to propagate the uncertainties in the deterministic models is a vital issue in the mechanical analysis of graphene. The Latin hypercube sampling (LHS) method is one of the advanced Monte Carlo simulations (MCS). The MCS has been applied in the investigation of the phase transition and magnetism of graphene sheets [14–17]. When the number of sampling is sufficient, it can reach a good accuracy in numerical computations. It is common to take the results of the MCS as an exact solution or comparison standard [18,19]. By dividing the sample spaces into subspaces, the LHS method effectively avoids the situation of point clustering together and repeating in the MCS [20].

The Kriging surrogate model (KSM) is an interpolation method that finds its roots in geostatistics [21,22]. With applications in the design of computer experiments [23], Bayesian prediction is used in the deterministic functions. Besides, global optimization is efficient in the expensive black-box functions [24]. Martin and Simpson [25] discussed the application of KSM to approximate deterministic models. Kleijnen [26] wrote the review to conclude the Kriging metamodeling in simulation. Wu [27] and Wu et al. [28] used the Kriging model in the inverse uncertainty quantification of nuclear reactor simulators under the Bayesian framework. Stein [29] explored the interpolation of spatial data for KSM. Besides, the introductions about KSM in statistics for spatial data were also published [30]. Moreover, Forrester and Keane [31] and Roustant et al. [32] developed the Kriging surrogate-based optimization. As one of the most promising spatial correlation models, the Kriging model is more flexible than the regression model and not as complicated and time-consuming as other metamodels [33]. The Kriging model is attractive for its prediction accuracy and time saving for the complicated analysis.

This chapter proposes the application of the KSM to represent the uncertain and complicated relationship between the elastic response of graphene sheets and the external forces. The LHS method is combined with the finite element model to successfully propagate the uncertainties in parameters corresponding to material and geometrical properties. The accuracy and convergence are confirmed by comparison with the reported references. The uncertainty analysis of zigzag and armchair graphene sheets in free vibration is completed and discussed.

10.2   MODEL FORMATION

10.2.1   GRAPHENE SHEETS

The modified Morse potential is successfully employed to simulate the nonlinear response of nanomaterials under tensile and torsional loading conditions [34]. The effects of defects on Young's modulus are studied by the modified Morse potential [35].

In a comparative study, the modified Morse potential provides more accurate predictions of tensile strength and fracture strain for carbon nanotubes than the reactive empirical bond order potential [36,37]. But in this study, the exact values of the material and geometrical parameters are represented by the corresponding interval ranges as shown in Table 10.1 according to the related references [38–46].

In the finite element model of graphene sheets, the length of the C–C bond corresponds to the length of the beam in the planar-frame structure, while the wall thickness is related to the diameter of the cross section in the beam elements. Figure 10.1 illustrates a schematic diagram of mono-layer graphene in two dimensions. The

---

TABLE 10.1

Geometrical and Material Parameters for Graphene Sheets

	Definition	Interval	Units
B_z	The length of bonds in the zigzag type	0.15–0.4	nm
Ba	The length of bonds in the armchair type	0.15–0.4	nm
Dz	The diameter of bond section in the zigzag type	0.02–0.05	nm
Da	The diameter of bond section in the armchair type	0.02–0.05	nm
Wz	The number of hexagons in width of the zigzag type	6–20	/
Wa	The number of hexagons in width of the armchair type	6–20	/
Hz	The number of hexagons in height in the zigzag type	20–60	/
Ha	The number of hexagons in height in the armchair type	20–60	/
Ez	Young's modulus of graphene sheets in the zigzag type	0.2–2	TPa
Ea	Young's modulus of graphene sheets in the armchair type	0.2–2	TPa
Rz	Poisson ratio of graphene sheets in the zigzag type	0.1–0.5	/
Ra	Poisson ratio of graphene sheets in the armchair type	0.1–0.5	/
Tz	Physical density of graphene sheets in the zigzag type	1500–4000	kg/m³
Ta	Physical density of graphene sheets in the armchair type	1500–4000	kg/m³

The way of probability distribution for each variable is uniform.

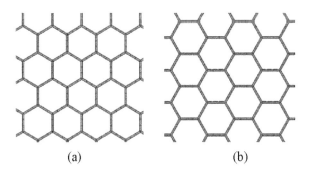

(a)   (b)

FIGURE 10.1 Graphene sheets with beam finite elements (a is for the zigzag type, and b is for the armchair type).

zigzag and armchair types are both included. The beam elements applied in graphene sheets are based on the Timoshenko beam theory as shown in the previous work [47].

However, the KSM proposed in this study for uncertainty analysis of graphene based on the finite element method is different from the Monte Carlo-based finite element method in the previous work. First, the motivation of KSM is for the uncertainty analysis of pristine graphene. The Monte Carlo-based finite element method (MC-FEM) is to propagate the vacancy defects in the pristine graphene. Second, the research objects are different. The MC-FEM is more concentrated in the geometrical uncertainties, namely vacancy defects. This study is more comprehensive to explore the uncertainties in the material and geometrical parameters. Third, compared with MC-FEM, the KSM has the following advantages: compatibility to the experimental data, convenience to the reliable prediction, and good performance in vibration analysis and uncertainty analysis for graphene. The MC-FEM is an effective numerical simulation to take the randomly distributed vacancy defects into consideration. However, the MC-FEM is not compatible with the experimental data or act as a surrogate model to represent the relationships between the input materials and geometrical parameters with the output resonant frequencies and their variance and fluctuations. Last, besides the vacancy defects in graphene, the uncertainties and fluctuations in the material and geometrical parameters are also important issues to deal with. The work in this study is a crucial supplement for the mechanical research of graphene.

10.2.2 LHS Method

The LHS method is one kind of advanced MCS. By dividing the range of each variable into disjoint intervals with equal probabilities, the samples are randomly selected from each interval in LHS. It improves the stability of MCS and keeps satisfying accuracy and good convergence. Consider a statistic system described by the function,

$$Y = F(X) \quad X = \{X_1, X_2, \ldots, X_n\} \tag{10.1}$$

where X is the random vector and represents the independent input random variables. F is the operator which performs computer simulation, such as the finite element computation.

The LHS method divides the range of each vector component into disjoint subsets with equal probabilities. Samples of each vector component are captured from the respective subsets according to Equation (10.1),

$$x_{ki}^j = P_{X_i}^{-1}(U_{ij}) \tag{10.2}$$

where $i = 1, \ldots, n$; $j = 1, \ldots, m$. n refers to the total number of vector components or dimensions of vector, and m is the number of subsets in a design. k is the subscript denoting a specific sample. P is the cumulative distribution function.

10.3   PROGRAM IMPLEMENTATION

To demonstrate the KSM for graphene sheets, Figure 10.2 depicts the flowchart of KSM, which can be clearly concluded in blocks distinguished by different colors.

The blue boxes represent the procedure of the deterministic FEM for vibration analysis of graphene sheets. First, the geometrical configuration of graphene sheets is defined; the corresponding parameters of bond length and sectional diameter, as well as the height and width of a hexagonal 2D lattice, are settled. Next, material parameters, which include Young's modulus, Poisson ratio, and physical density, are provided. Then, FEM was performed to compute the vibration modes and related natural frequencies.

After validation of the FEM, the loop performs until sufficient times of sampling are completed as presented in the red boxes. In the continuous loop, parameters corresponding to geometrical and material properties are the input variables of KSM, while the natural frequencies of the free vibration by finite element analysis are also captured and transferred to the procedure of KSM. These two groups of databases form sampling pairs in KSM.

In the pink boxes, the input variables of LHS and the output results of the finite element method (FEM) are sample pairs in KSM. To obtain satisfying prediction accuracy and good convergence, regression and correlation functions are required to determine. The prediction results of KSM can be confirmed by comparison with the results in the reported references.

The numerical simulation model of graphene sheets was created by the ANSYS Parameter Design Language (Version 14.5, APDL, ANSYS, USA). Carbon atoms in graphene are bonded together with covalent bonds forming a hexagonal 2D lattice.

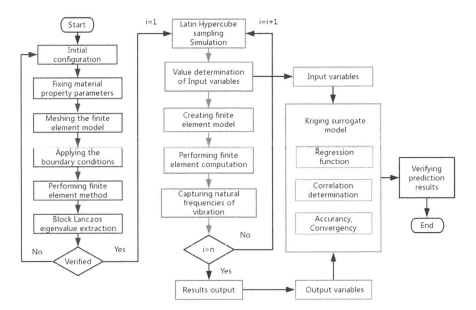

FIGURE 10.2   The flowchart of the Kriging surrogate model.

The displacement of individual atoms under an external force is constrained by the bonds. For the modeling of the C–C bonds, the 3D elastic BEAM188 element was used. Each node has six degrees of freedom, and the translations and rotations around the x, y, and z directions are included. In Table 10.1, the corresponding parameters which are related to the geometrical and material properties are listed. The interval ranges of material and geometrical parameters are settled according to the relevant experimental and numerical data in the literature [8,48,49]. The exact values of the parameters are replaced by the corresponding wide intervals.

10.4   DISCUSSION AND RESULTS

In the deterministic beam FEM, there are 6226 beams with 16,664 nodes. By LHS, the stochastic values for each parameter in certain intervals are sampled for the FEM of graphene sheets.

10.4.1   STATISTICAL RESULTS

As mentioned above, the uncertainties in the geometrical and material properties are propagated in the FEM by LHS. In the computational process, 1000 samples for each input parameter and output result are obtained. Based on the original database, the results are studied by stochastic mathematics and the probability theory as shown in Table 10.2 and Figures 10.3–10.5.

In Table 10.2 and Figure 10.3, the differences between the zigzag and armchair graphene sheets are observed. In general, the mean values of the first four-order natural frequencies are approximated in the zigzag and armchair graphene sheets. However, armchair graphene sheets have larger values in the variance, maximum, and minimum from the respective database. Especially under the minimum extreme condition, the results of armchair graphene sheets are evidently larger than those of the zigzag type. On the one hand, the difference in the geometrical structures of

---

TABLE 10.2
Probability Results of LHS in the Finite Element Model of Graphene

	Mean (THz)	Variance (THz$^{\wedge 2}$)	Maximum (THz)	Minimum (THz)
F_1-Z	3.0060	7.2567	21.1325	0.1654
F_2-Z	4.7177	18.3757	42.7713	0.2730
F_3-Z	6.2362	31.1114	44.2943	0.3413
F_4-Z	7.8909	49.5804	63.4948	0.4411
F_1-A	3.1309	9.7822	24.3613	0.2442
F_2-A	4.8098	21.6774	43.1523	0.4345
F_3-A	6.4075	38.0443	55.5136	0.5552
F_4-A	8.0964	62.0176	72.8086	0.7306

F_1–F_4 represent the first-fourth natural frequencies. Z and A represent zigzag and armchair graphene sheets, respectively.

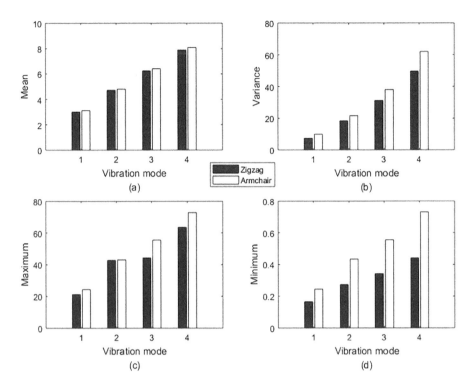

FIGURE 10.3 Probability results of LHS in the finite element model of graphene (a represents mean values, b represents variance values, c represents maximum values, and d represents minimum values; the unit for the natural frequency is THz).

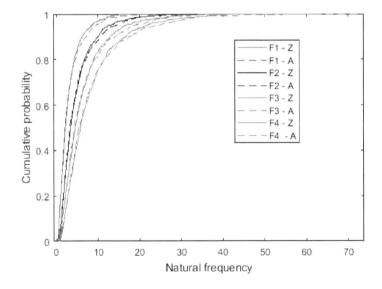

FIGURE 10.4 Cumulative probability of natural frequencies (the unit for the natural frequency is THz).

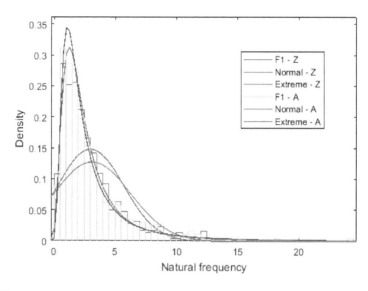

FIGURE 10.5   Probability density distribution for the first-order natural frequencies (the unit for the natural frequency is THz).

the armchair and zigzag graphene sheets causes divergence in the statistical results. On the other hand, the stochastic sampling process can also lead to the deviation in the database of natural frequencies of armchair and zigzag graphene sheets. Furthermore, the differences between the zigzag and armchair edges in natural frequencies are also caused by the boundary conditions. Even though the microstructures of the zigzag and armchair graphene have only a 90° rotation discrepancy, the boundary condition differences in the zigzag and armchair edges cannot be ignored. The six degrees of freedom (displacement in the x, y, and z axes and rotation around the x, y, and z axes) for atoms in the edges of graphene are supposed to be zero. The differences in zigzag and armchair edges lead to inequality in natural frequencies.

The cumulative probability of the first four-order natural frequencies is compared in Figure 10.4. For each order free vibration, the cumulative probability of the zigzag and armchair is close to each other. However, for all the first four-order natural frequencies both in zigzag and armchair graphene sheets, the drag section in the large probability approaching one is very long and not negligible. This phenomenon is also presented in the results of the probability density distribution in Figure 10.5.

Even though all the input variables (parameters corresponding to the geometrical and material properties) of LHS are uniformly distributed in the sample space, the output results (natural frequencies of graphene sheets) do not obey the uniform or strict normal distribution as shown in Figure 10.5. The drag section is an inevitable component in all ranges of natural frequencies. By taking comprehensive uncertainties in graphene sheets into consideration, the shape of the probability distribution of natural frequencies is the peak shape with a long drag, while the natural frequencies are the ratio of stiffness matrices and mass matrices. It is reasonable to explain that the test equivalent results are concentrated in the narrow intervals with the possibility of large extreme values.

10.4.2 Comparison and Discussion

The KSM is applied to predict the natural frequencies of graphene sheets. A comparison between the prediction results of KSM and those of the reported references [41–49] is demonstrated in Figures 10.6–10.8.

In Figure 10.6, the prediction results of KSM are compared with those of Liu et al. [38], Kudin et al. [39], and Wei et al. [40]. In the first-order natural frequency, the prediction results of KSM are larger than the results of Liu et al. [38], Kudin et al. [39], and Wei et al. [40]. The armchair graphene sheets have closer results to the reported reference [41–46]. In the second order, the prediction results are smaller than the reported results. The zigzag graphene sheets have the more approximated results. The prediction results of zigzag graphene sheets have satisfied agreement with the reported results in the third-order natural frequency. In the fourth-order natural frequency, the zigzag-type graphene sheets have the more consistent results. In addition, in the first four-order natural frequencies, the prediction results of zigzag graphene sheets are all larger than the armchair graphene sheets. Furthermore, to fit the relationship between the natural frequencies and parameters corresponding to the geometrical and material properties, the constant, linear, and quadratic functions in KSM are applied and compared in Figure 10.6. It is obvious to find that besides the accuracy, the satisfactory convergency is reached in the different orders of KSM.

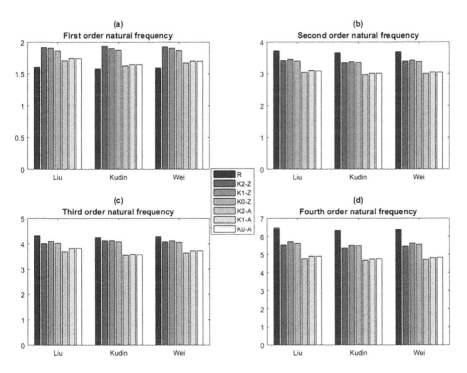

FIGURE 10.6 (a–d) Comparison between the prediction results of KSM and that of DFT (the unit for the natural frequency is THz; K0–K2 are for the constant, linear, and quadratic functions; and Z and A are for zigzag and armchair graphene, respectively).

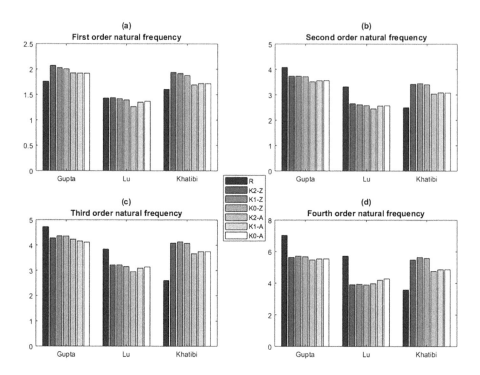

FIGURE 10.7 (a–d) Comparison between the prediction results of KSM and those of MD (the units for the natural frequency is THz; K0–K2 are for constant, linear, and quadratic functions; and Z and A are for zigzag and armchair graphene sheets, respectively).

The prediction results of KSM are all larger than that of Sadeghzadeh and Khatibi [43] in the first four-order natural frequencies of graphene sheets as shown in Figure 10.7. Moreover, the zigzag graphene sheets have larger natural frequencies than armchair graphene sheets. However, when the parameters corresponding to the geometrical and material properties take the same values of Gupta et al. [41] and Lu and Huang [42], the difference between zigzag and armchair graphene sheets is not evident. In addition, the prediction results of KSM in the first order are larger than that of Gupta et al. [41] but smaller in the other three order natural frequencies. The prediction results of KSM in the first-order natural frequency are close to the results of Lu and Huang [42], especially in the zigzag graphene sheets. But the prediction results of KSM are smaller than that of Lu and Huang [42] in the other three order natural frequencies. In Figure 10.7, the convergence of KSM in distinct orders is also proved.

In Figure 10.8, for the first-order natural frequency, the prediction results of KSM have appropriate accuracy with that of Reddy et al. [45] and Zhou et al. [46]. Evident deviation in the prediction results of KSM with that of Cadelano et al. [44] is observed in the first four-order natural frequencies. In addition, the prediction results are smaller than that of Reddy et al. [45] and Zhou et al. [46]. The difference between the zigzag and armchair is not negligible in the situation of Reddy et al. [45] and Zhou et al. [46]. Even though there is a deviation between the prediction results

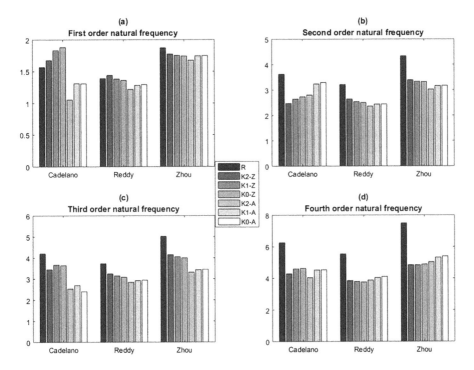

FIGURE 10.8   (a–d) Comparison between the prediction results of KSM and that of MM (the unit for the natural frequency is THz; K0–K2 are for the constant, linear, and quadratic functions, and Z and A represent for zigzag and armchair graphene sheets, respectively).

of KSM and that of Reddy et al. [45], Zhou et al. [46], and Cadelano et al. [44], the convergence in different orders of KSM demonstrates the robustness and reliability of KSM.

In addition, the results in this study are in good agreement with those of the nonlocal plate model proposed by Ansari et al. [50]. Different from the Timoshenko beam FEM, the Mindlin plane equations are coupled with van der Waals interaction based on the nonlocal constitutive elastic equations. However, the out-of-plane behavior of graphene in the beam FEM is larger than the results in the real situation due to the bond angle bending interaction. Besides, the geometrical parameters, such as the diameters, chiral angles, and length of carbon bonds, have crucial influences on the axial and shear deformation coupling behaviors [51] and buckling critical stress [52]. More appropriate theoretical models for the analysis of mechanical properties of graphene are necessary to be explored [53].

10.4.3   Uncertainty Analysis

The competitive prediction competence in the KSM not only is applied in deterministic models but also can be used in uncertainty analysis. Based on the satisfying accuracy and robust convergence, KSM is applied in the uncertainty analysis of geometrical and material properties in graphene sheets.

Tables 10.3 and 10.4 list the results of uncertainty analysis of geometrical and material properties, respectively. When the corresponding parameters are sampled following a uniform distribution from the specific interval ranges, the probability density distributions of the first-order natural frequency are demonstrated in Figures 10.9 and 10.10.

In Figure 10.9a, when the length of bonds in graphene sheets is uncertain, the probability density distribution of zigzag and armchair graphene sheets is contiguous. The situation is analogous for the number of hexagons in height as presented in Figure 10.10d. However, for the diameter of the bond section and the number of hexagons in width, the difference in the probability density distribution between zigzag and armchair graphene sheets is apparent. In general, the uncertainties in the diameter of bond section and the number of hexagons in width lead to the lower and more gentle probability distribution in the armchair than that in the zigzag graphene sheets as shown in Figure 10.9b and c. The zigzag graphene sheets have a more concentrated and peak probability density distribution in the more narrow result interval. Besides, in Table 10.3, the values of variance, maximum, and minimum also confirm this point. In a sense, zigzag graphene sheets are more robust and less sensitive to the

TABLE 10.3

Uncertainty Analysis about Parameters of Geometrical Properties

	Interval	Mean (THz)	Variance (THz^2)	Maximum (THz)	Minimum (THz)
Bz (nm)	0.2–0.35	2.1151	0.0367	2.3469	1.7151
Ba (nm)	0.2–0.35	2.1472	0.1742	2.6792	1.3697
Dz (nm)	0.025–0.045	2.0073	0.2933	3.1165	1.0013
Da (nm)	0.025–0.045	1.9239	0.3184	3.0748	1.0522
Wz	8–18	1.9574	0.1890	2.9359	1.2455
Wa	8–18	2.0740	0.4036	3.7651	1.3056
Hz	30–50	2.0945	0.0695	2.7333	1.6979
Ha	30–50	1.9636	0.0796	2.5998	1.4943

TABLE 10.4

Uncertainty Analysis about Parameters of Material Properties

	Interval	Mean (THz)	Variance (THz^2)	Maximum (THz)	Minimum (THz)
Ez (TPa)	0.6–1.3	1.7847	0.0457	2.1255	1.3774
Ea (TPa)	0.6–1.3	1.5794	0.0639	2.0562	1.1876
Rz	0.16–0.3	1.9990	0.0016	2.0442	1.9204
Ra	0.16–0.3	1.9200	0.0004	1.9421	1.8707
Tz (g/cm^3)	1.6–3.6	2.0531	0.1223	2.5258	1.4059
Ta (g/cm^3)	1.6–3.6	1.9587	0.0583	2.3490	1.5493

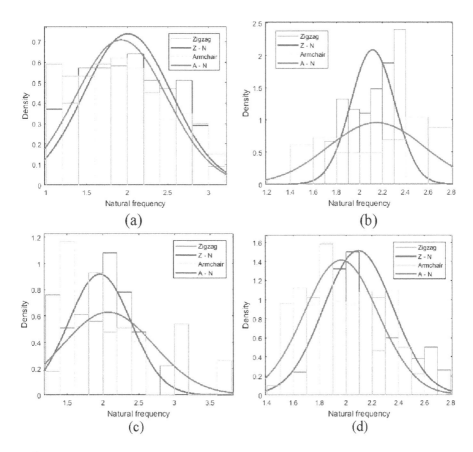

FIGURE 10.9 Probability density distribution of natural frequency for uncertainty analysis (a is for Bz and Ba; b is for Dz and Da; c is for Wz and Wa; d is for Hz and Ha; the unit for the natural frequency is THz).

uncertainties in the diameter of bond section and the number of hexagons in width than armchair graphene sheets.

In addition, for the uncertainties in material properties in Figure 10.11, the probability density distribution of zigzag graphene sheets is on the right side of armchair graphene sheets, especially when the Poisson ratio is random. In Table 10.5, the mean values of the first-order natural frequency in zigzag graphene sheets are all larger than that of armchair graphene sheets when Young's modulus, Poisson's ratio, or mass density are stochastic and uncertain. The variances of zigzag graphene sheets are also larger than that of armchair graphene sheets. In the free vibration, the uncertainties in material properties lead to more evident fluctuation in zigzag graphene sheets.

Compared with the FEM computation, the advantage of time saving in the KSM is presented in the uncertainty analysis process of graphene sheets. The time of natural frequencies and vibration mode calculation for each deterministic sample of graphene by FEMs is nearly 5.6 seconds. However, based on the original database

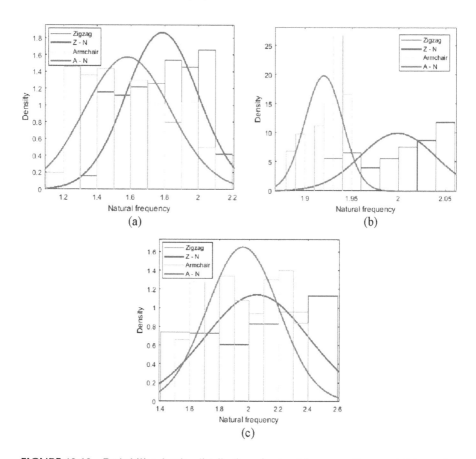

FIGURE 10.10   Probability density distribution of natural frequency for uncertainty analysis (a is for Ez and Ea; b is for Rz and Ra; c is for Tz and Ta; the unit for the natural frequency is THz).

of LHS from the FEM of graphene, the KSM can be applied to predict the natural frequencies with high efficiency. For the prediction of 1000 samples of graphene, it only takes approximately 12.3 seconds by KSM. When the parameters corresponding to geometrical and material properties are randomly distributed, performing the FEM for sufficient times is computationally expensive and time-consuming. Thus, the advantage of time saving in the KSM compared with the traditional FEM is evident.

In addition, even when the exact mathematical relationship or functional expression between the graphene size and natural frequencies is not explicit, the KSM is proposed to effectively describe the implicit relationship. Furthermore, the KSM is compatible with the database of the numerical simulation and experimental tests. It is possible to combine the results of numerical simulation and physical experiments and create a more inclusive and smart database. Therefore, developing the KSM for the study of graphene sheets and related nanomaterials is promising and crucial.

10.5 CONCLUSION

In summary, the KSM is not only a powerful method with satisfying accuracy and convergence in prediction, but also convenient and time saving in uncertainty analysis of graphene sheets. The difficulties in uncertainty and vibration analysis of graphene sheets are well overcome by the KSM based on the LHS. With the comprehensive uncertainties in geometrical and material properties, the probability results of the LHS illustrate that the concentrated narrow interval with a long drag section is more appropriate than the strict normal probability density distribution. Furthermore, zigzag graphene sheets are more robust and less sensitive to the uncertainties of the geometrical property than armchair graphene sheets in the free vibration. However, the uncertainties in material properties cause a larger fluctuation in zigzag graphene sheets than that in armchair graphene sheets.

REFERENCES

1. Geim, A.K., Novoselov, K.S. The rise of graphene. *Nature Materials*, **2007**, 63, 183–191.
2. Novoselov, K.S., et al. Electric field effect in atomically thin carbon films. *Science*, **2004**, 306(5696), 666–669.
3. Novoselov, K.S., et al. Two-dimensional gas of massless Dirac fermions in graphene. *Nature*, **2005**, 4387065, 197.
4. Zhang, Y., et al. Experimental observation of the quantum Hall effect and Berry's phase in graphene. *Nature*, **2005**, 4387065, 201.
5. Balandin, A.A., et al. Superior thermal conductivity of single-layer graphene. *Nano Letters*, **2008**, *83*, 902–907.
6. Ghosh, D., et al. Extremely high thermal conductivity of graphene: Prospects for thermal management applications in nanoelectronic circuits. *Applied Physics Letters*, **2008**, 9215, 151911.
7. Calizo, I., et al. Temperature dependence of the Raman spectra of graphene and graphene multilayers. *Nano Letters*, **2007**, 79, 2645–2649.
8. Lee, C., et al. Measurement of the elastic properties and intrinsic strength of monolayer graphene. *Science*, **2008**, 321(5887), 385–388.
9. Blakslee, O.L., et al. Elastic constants of compression-annealed pyrolytic graphite. *Journal of Applied Physics*, **1970**, 418, 3373–3382.
10. Yu, M., et al. Strength and breaking mechanism of multiwalled carbon nanotubes under tensile load. *Science*, **2000**, 2875453, 637–640.
11. Poot, M., Herre, Z.H.S.J. Nanomechanical properties of few-layer graphene membranes. *Applied Physics Letters*, **2008**, 926, 063111.
12. Frank, I.W., et al. Mechanical properties of suspended graphene sheets. *Journal of Vacuum Science & Technology B: Microelectronics and Nanometer Structures Processing Measurement and Phenomena*, **2007**, 256, 2558–2561.
13. Georgantzinos, S.K., Markolefas, S., Giannopoulos, G.I., et al. Designing pinhole vacancies in graphene towards functionalization: Effects on critical buckling load. *Superlattices and Microstructures*, **2017**, 103, 343–357.
14. Ulybyshev, M.V., et al. Monte Carlo study of the semimetal-insulator phase transition in monolayer graphene with a realistic interelectron interaction potential. *Physical Review Letters*, **2013**, 1115, 056801.
15. Armour, W., Hands, S., Strouthos, C. Monte Carlo simulation of the semimetal-insulator phase transition in monolayer graphene. *Physical Review B*, **2010**, 8112, 125105.

16. Feldner, H., et al. Magnetism of finite graphene samples: Mean-field theory compared with exact diagonalization and quantum Monte Carlo simulations. *Physical Review B*, **2010**, 8111, 115416.

17. Whitesides, R., Frenklach, M. Detailed kinetic Monte Carlo simulations of graphene-edge growth. *The Journal of Physical Chemistry A*, **2009**, 1142, 689–703.

18. Chu, L., et al. Application of Latin hypercube sampling based kriging surrogate models in reliability assessment. *Science Journal of Applied Mathematics and Statistics*, **2015**, 3, 263–274.

19. Chu, L., et al. Reliability based optimization with metaheuristic algorithms and Latin hypercube sampling based surrogate models. *Applied and Computational Mathematics*, **2015**, 4, 462–468.

20. Helton, J.C., Davis, F.J. Latin hypercube sampling and the propagation of uncertainties in analyses of complex systems. *Reliability Engineering & System Safety*, **2003**, 811, 23–69.

21. Matheron, G. Principles of geostatistics. *Economic Geology*, **1963**, 58, 1246–1266.

22. Cressie, N. The origins of kriging. *Mathematical Geology*, **1990**, 22(3), 239–252.

23. Currin, C., Mitchell, T., Morris, M., Ylvisaker, D. Bayesian prediction of deterministic functions, with applications to the design and analysis of computer experiments. *Journal of the American Statistical Association*, **1991**, 86(416), 953–963.

24. Jones, D.R., Schonlau, M., Welch, W.J. Efficient global optimization of expensive black-box functions. *Journal of Global optimization*, **1998**, 13(4), 455–492.

25. Martin, J.D., Simpson, T.W. Use of kriging models to approximate deterministic computer models. *AIAA Journal*, **2005**, 43(4), 853–863.

26. Kleijnen, J.P. Kriging metamodeling in simulation: A review. *European Journal of Operational Research*, **2009**, 192(3), 707–716.

27. Wu, X. Metamodel-based inverse uncertainty quantification of nuclear reactor simulators under the Bayesian framework. Doctoral dissertation, University of Illinois at Urbana-Champaign, 2017.

28. Wu, X., Kozlowski, T., Meidani, H. Kriging-based inverse uncertainty quantification of nuclear fuel performance code BISON fission gas release model using time series measurement data. *Reliability Engineering & System Safety*, **2018**, 169, 422–436.

29. Stein, M.L. *Interpolation of Spatial Data: Some Theory for Kriging*, Springer Science & Business Media. 2012.

30. Cressie, N. *Statistics for Spatial Data*, revised edition, John Wiley & Sons. 2015.

31. Forrester, A.I., Keane, A.J. Recent advances in surrogate-based optimization. *Progress in Aerospace Sciences*, **2009**, 45(1), 50–79.

32. Roustant, O., Ginsbourger, D., Deville, Y. Dicekriging, diceoptim: Two r packages for the analysis of computer experiments by kriging-based metamodelling and optimization. *Journal of Statistical Software*, **2012**, 51(1), 54.

33. Echard, B., Gayton, N., Lemaire, M. AK-MCS: An active learning reliability method combining Kriging and Monte Carlo simulation. *Structural Safety*, **2011**, 332, 145–154.

34. Wernik, J.M., Meguid, S.A. Atomistic-based continuum modeling of the nonlinear behavior of carbon nanotubes. *Acta Mechanica*, **2010**, 2121(2), 167–179.

35. Parvaneh, V., Shariati, M. Effect of defects and loading on prediction of Young's modulus of SWCNTs. *Acta Mechanica*, **2011**, 2161(4), 281–289.

36. Brenner, D.W., et al. A second-generation reactive empirical bond order REBO potential energy expression for hydrocarbons. *Journal of Physics: Condensed Matter*, **2002**, 144, 783.

37. Duan, W.H., et al. Molecular mechanics modeling of carbon nanotube fracture. *Carbon*, **2007**, 459, 1769–1776.

38. Liu, F. Ming, P., Li, J. Ab initio calculation of ideal strength and phonon instability of graphene under tension. *Physical Review B*, **2007**, 766, 064120.

39. Kudin, K.N., Scuseria, G.E., Yakobson, B.I. C 2 F BN and C nanoshell elasticity from ab initio computations. *Physical Review B*, **2001**, 6423, 235406.
40. Wei, X., et al. Nonlinear elastic behavior of graphene: Ab initio calculations to continuum description. *Physical Review B*, **2009**, 8020, 205407.
41. Gupta, S., Dharamvir, K., Jindal, V.K. Elastic moduli of single-walled carbon nanotubes and their ropes. *Physical Review B*, **2005**, 7216, 165428.
42. Lu, Q., Huang, R. Nonlinear mechanics of single-atomic-layer graphene sheets. *International Journal of Applied Mechanics*, **2009**, 103, 443–467.
43. Sadeghzadeh, S., Khatibi, M.M. Modal identification of single layer graphene nano sheets from ambient responses using frequency domain decomposition. *European Journal of Mechanics - A/Solids*, **2017**, 65, 70–78.
44. Cadelano, E., et al. Nonlinear elasticity of monolayer graphene. *Physical Review Letters*, **2009**, 10223, 235502.
45. Reddy, C.D., Rajendran, S., Liew, K.M. Equilibrium configuration and continuum elastic properties of finite sized graphene. *Nanotechnology*, **2006**, 173, 864.
46. Zhou, L., Wang, Y., Cao, G. Elastic properties of monolayer graphene with different chiralities. *Journal of Physics: Condensed Matter*, **2013**, 25(12), 125302.
47. Chu L., Shi J.J., Souza de Cursi, E. Vibration analysis of vacancy defected graphene sheets by Monte Carlo based finite element method. *Nanomaterials*, **2018**, 8(7), 489.
48. Warner, J.H., Lee, G.D., He, K., Robertson, A.W., et al. Bond length and charge density variations within extended arm chair defects in graphene. *ACS Nano*, **2013**, 7(11), 9860–9866.
49. Fasolino, A., Los, J.H., Katsnelson, M.I. Intrinsic ripples in graphene. *Nature Materials*, **2007**, 6(11), 858.
50. Ansari, R., Rajabiehfard, R., Arash, B. Nonlocal finite element model for vibrations of embedded multi-layered graphene sheets. *Computational Materials Science*, **2010**, 49(4), 831–838.
51. Georgantzinos, S.K., Giannopoulos, G.I., Anifantis, N.K. On the coupling of axial and shear deformations of single-walled carbon nanotubes and graphene: A numerical study. *Proceedings of the Institution of Mechanical Engineers, Part N: Journal of Nanoengineering and Nanosystems*, **2010**, 224(4), 163–172.
52. Rouhi, S., Ansari, R. Atomistic finite element model for axial buckling and vibration analysis of single-layered graphene sheets. *Physica E: Low-dimensional Systems and Nanostructures*, **2012**, 44(4), 764–772.
53. Georgantzinos, S.K. A new finite element for an efficient mechanical analysis of graphene structures using computer aided design/computer aided engineering techniques. *Journal of Computational and Theoretical Nanoscience*, **2017**, 14(11), 5347–5354.

11 Equivalent Young's Modulus Prediction

11.1 INTRODUCTION

In graphene, an extraordinary intrinsic in-plane strength is observed in experiments and predicted in analytical theories. However, the reported Young's modulus of graphene is oscillated in a large interval and has not been settled as a certain value. Lee et al. [1] measured Young's modulus to be 1.0 TPa by atomic force microscopy (AFM). Yanovsky et al. [2] used the quantum mechanics and the predicted Young's modulus is 0.737 TPa. By molecular dynamics (MD) simulation, Ni et al. [3] demonstrated Young's modulus to be 1.1 TPa, Tsai and Tu [4] reported Young's modulus to be 0.912 TPa, and Ansari et al. [5] demonstrated Young's modulus to be 0.8 TPa. Besides, more extensive investigation in MD is attempted by Zhang et al. [6] and Javvaji et al. [7]. Moreover, based on Brenner's potential, Reddy et al. [8] reported Young's modulus to be 1.095–1.125 and 1.106–1.201 TPa for armchair and zigzag graphene sheets, respectively. Georgantzinos et al. [9] explored the finite element method (FEM) and expressed the value of Young's modulus to be 1.367 TPa. Furthermore, with truss-type analytical models, Scarpa et al. record 1.042 and 1.040 TPa for armchair and zigzag sheets, respectively [10]. Shokrieh and Rafiee [11] developed a honeycomb-like discrete structure and the achieved Young's modulus is 1.04 TPa.

The deviation and fluctuation in the analytical and measured results of Young's modulus of graphene are mainly caused by the intrinsic and inescapable uncertainty or defects in the microstructure. The investigation of the influence of defects in graphene is an essential and challenging issue. Defects are unpredictable and randomly distributed in the entire graphene lattice. It is hard to control the size and placement of defects in the formation process of chemical vapor deposition [12], chemical reduction of graphene oxides [13], and mechanical exfoliation [14].

More attention is paid to the investigation of vacancy defected graphene. Zandiatashbar et al. [15] applied AFM nanoindentation to quantify the stiffness and strength of defected graphene. The disordered structure of graphene is observed by Raman spectroscopy [16,17]. Qin et al. [18] found that wrinkled multilayer graphene has geometrical locking effects, which contribute to the enhancement effects. However, Min and Aluru [19] stated that the fracture stress is lower in winkled graphene (60 GPa) than that in flat graphene (97.5 GPa) in the shear test on zigzag graphene by MD simulations. Papageorgiou et al. [20] reported that with the sp³-defects in graphene, the strength and stiffness are still maintained even when the defect percentage is high, while vacancy defects in graphene cut down the strength abruptly. Lópezpolín et al. [21] attempted experiments of controlled vacancy defects

DOI: 10.1201/9781003226628-13

in graphene. Specifically, Young's modulus raises with the augment of vacancy defect amounts and reaches a peak when the percentage of vacancy defects is 0.2%.

The uncertainty and complicity in microstructure graphene lead to the appearance of contradictory results or phenomena. It is meaningful to explore the impacts of vacancy defects in graphene. In the present study, the finite element model is developed based on the molecular mechanics theory. The equivalent Young's modulus is computed from the elastic strain energy. The impacts of vacancy defects in graphene are discussed by two different patterns: the regular deterministic location and random distribution. To describe the random vacancy defects, the Monte Carlo-based finite element method (MC-FEM) is applied to propagate the stochastic dispersion of vacancy defects in the entire graphene.

11.2   MATERIALS AND METHODS

With carbon atoms arranged in a honeycomb microstructure, graphene can be simplified as a triangular lattice with a basis of two atoms per unit cell. The vectors in the lattice can be expressed as (Figure 11.1)

$$\mathbf{a}_1 = \frac{a}{2}(3, \sqrt{3}), \quad \mathbf{a}_2 = \frac{a}{2}(3, -\sqrt{3}) \tag{11.1}$$

where $a \approx 1.42 \overset{\circ}{A}$ is the distance between neighbor carbon atoms. The reciprocal lattice vectors are given by

$$\mathbf{b}_1 = \frac{2\pi}{3a}(1, \sqrt{3}), \quad \mathbf{b}_2 = \frac{2\pi}{3a}(1, -\sqrt{3}) \tag{11.2}$$

The three nearest neighbor vectors can be written as

$$\delta_1 = \frac{a}{2}(1, \sqrt{3}), \quad \delta_2 = \frac{a}{2}(1, -\sqrt{3}), \quad \delta_3 = -a(1,0) \tag{11.3}$$

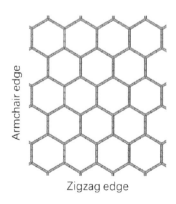

FIGURE 11.1   Honeycomb lattice of graphene sheets.

The modified Morse potential [22] can be written as

$$U_{\text{Morse}} = U_r + U_\theta \tag{11.4}$$

The analytical function representing the bond energy is the Morse function:

$$U_r = D_{ij}^e \left[e^{(-2a_{ij}\Delta r_{ij})} - 2e^{(-a_{ij}\Delta r_{ij})} \right] \tag{11.5}$$

where D_{ij}^e is the bond stretching energy, r_{ij} is the equilibrium distance, Δr_{ij} represents the variation of the bond length, and a_{ij} is a relative coefficient. With the parameter for hybridized sp^2 bonds, the Morse potential is expressed as

$$U_r = D_e \left\{ \left[1 - e^{-\beta(r-r_0)} \right]^2 - 1 \right\} \tag{11.6}$$

where r_0 is the bond equilibrium length, D_e is the energy of dissociation, and β is the coefficient of regression fitting.

$$r_0 = 0.139 \text{ nm}, D_e = 6.03105 \times 10^{-10} \text{ Nnm}, \beta = 2.625 \times 10^{10} \text{ m}^{-1} = 26.25 \text{ nm}^{-1} \tag{11.7}$$

The energy of bond angle is written as

$$U_\theta = \frac{1}{2} k_\theta (\Delta\theta)^2 [1 + k_{\text{sextic}} (\Delta\theta)^4] \tag{11.8}$$

with $k_\theta = 0.9 \times 10^{-18}$ Nm/rad^2, $\Delta\theta = \theta - \theta_0$, $\theta_0 = 2.094$ rad, $k_{\text{sextic}} = 0.754$ rad^{-4}.

For axial tension, differentiation of the stretching potential can be obtained:

$$F(\Delta r) = 2\beta D_e (1 - e^{-\beta\Delta r}) e^{-\beta\Delta r} \tag{11.9}$$

With the finite element, the bond stretching and angle variations can be successfully simulated [23,24]. For the vacancy-defected graphene under axial tension, the equivalent Young's modulus can be computed by

$$\tilde{E} = \frac{1}{V_0} \left(\frac{\partial^2 U_B}{\partial \varepsilon^2} \right)_{\varepsilon=0} \tag{11.10}$$

where V_0 is the effective volume of graphene, U_B is the total strain energy, and ε is the tensile strain. The thickness of graphene is not deterministic and usually settled as 0.34 nm. The density of strain energy D_B is more convenient in analysis.

$$D_B = \frac{1}{2} \tilde{E} \varepsilon^2 \tag{11.11}$$

For the finite element model, the equivalent Young's modulus is written as,

$$\tilde{E} = 2 \sum_{i=1}^{i=n} D_{Bi} / \varepsilon_i^2 \tag{11.12}$$

where n is the number of finite elements in the graphene lattice, D_{Bi} is the strain energy density of the i th element, and ε_i is the elastic strain of the i th element.

The Poisson ratio can be defined as

$$v = -\frac{\Delta w}{w} \bigg/ \frac{\Delta L}{L} \tag{11.13}$$

where w is the width of the graphene sheet, L is the length, and Δw and ΔL are the changes in the width and length, respectively. For a material undergoing a shear deformation, G is similarly defined as

$$G = \frac{1}{V_0}\left(\frac{\partial^2 U_B}{\partial \phi^2}\right)_{\phi=0} \tag{11.14}$$

11.3   RESULTS AND DISCUSSION

The finite element model of pristine graphene is presented in Figure 11.2, which is based on the previous work of this study [25]. For each node, there are six degrees of freedom, namely displacement and rotation in the X, Y, and Z axes. The shear stress is loaded in the armchair edge and zigzag edge, respectively. The contour results of vector sum displacement are also depicted in Figure 11.2 for pristine graphene. Young's modulus in the pristine graphene is 1.2 TPa. To discuss the effects of vacancy defects on the mechanical properties of graphene, the regular deterministic and randomly distributed vacancy defects are presented and discussed in the following.

11.3.1   REGULAR DETERMINISTIC VACANCY DEFECTS

The concentrated vacancy defects in the center of graphene are gradually amplified as shown in Figure 11.3. In Figure 11.4, according to the size amplification of the center-concentrated vacancy defects, the maximum displacement, energy density, and total strain in the entire graphene all increase when shear stress is loaded in the zigzag and armchair edges. In particular, the total strain in the entire graphene appears a fluctuation at the second size of center-concentrated vacancy defect when shear

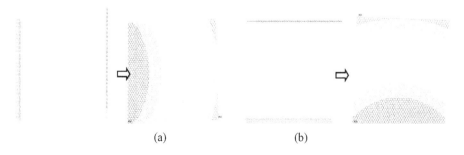

(a)   (b)

FIGURE 11.2   Finite element model of pristine graphene under shear stress (a is for shear stress in the armchair edge and b is for shear stress in the zigzag edge).

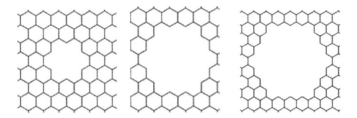

FIGURE 11.3 Center-concentrated vacancy defects.

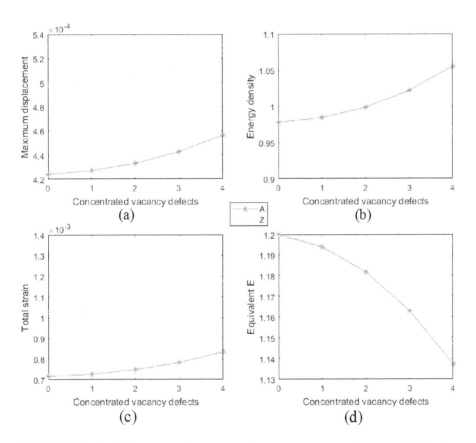

FIGURE 11.4 (a–d) The results of graphene with center-concentrated vacancy defects (the units for a, b, and d are m, J/m³, and TPa, respectively).

stress is in the zigzag edge. The maximum displacement and total strain of graphene under shear stress in the zigzag edge are larger than those under shear stress in the armchair edge, while in the energy density, the compared results between two different boundary conditions are contrary. Furthermore, the equivalent Young's modulus decreases with the enlargement of center-concentrated vacancy defects in both situations under shear stress in zigzag and armchair edges as shown in Figure 11.4d.

The center-concentrated vacancy defects cause evidently deterioration of elastic stiffness in graphene. This result reaches a good agreement with the reported literature [26,27]. The reduction of equivalent Young's modulus in graphene under shear stress in armchair stress is sharper than that in the zigzag edge. The difference caused by chirality in graphene cannot be negligible.

In addition, the vector sum of the displacement in graphene is demonstrated in Figures 11.5 and 11.6. The displacement around the center-concentrated vacancy defects is amplified when the defect size increases, especially in the edges of vacancy defects close to the loaded edge of graphene. More importantly, even though the situation under shear stress in the zigzag edge has larger maximum displacement in the entire graphene with the same center-concentrated vacancy defects, the reduction of Young's modulus in the situation under shear stress in the armchair edge is larger.

The vacancy defects in graphene are unpredictable in terms of placement and shape. For the deterministic regular vacancy defects, the above-mentioned center-concentrated vacancy is one of the typical kinds of defects. To evaluate the effects of vacancy defects in graphene, periodic and regular vacancy defects are introduced in one-quarter of the entire graphene as shown in Figure 11.7.

Even though the vacancy defects are regularly and periodically distributed in one-quarter of the entire graphene, the results in the maximum displacement, energy density, and total strain are similar to those of center-concentrated vacancy defects. Along with the increase of the number of vacancy defects, the maximum displacement, energy density, and total strain in the entire graphene are all magnified in both situations (shear stress in the armchair edge and in the zigzag edge), as shown in Figure 11.8. Graphene under shear stress in the zigzag edge has greater maximum displacement and total strain, but smaller energy density than that under shear stress in the armchair edge. Besides, there is no abrupt fluctuation in the results of maximum displacement, energy density, and total strain, and the approximated parallel tendency is observed in graphene under the two different boundary conditions.

However, for the equivalent Young's modulus of vacancy-defected graphene, the results of graphene under two different boundary conditions are distinct and non-parallel as shown in Figure 11.8d. Young's modulus of vacancy-defected graphene under shear stress in the armchair edge is quickly cut down according to the rise of the number of periodic vacancy defects. The results in graphene under shear stress in the zigzag edge are lightly and slowly decreased. The equivalent Young's modulus in graphene under shear stress in the zigzag edge is higher than that in the armchair edge. In general, the equivalent Young's modulus is sensitive to the boundary condition, but graphene under shear stress in the zigzag edge is more robust to defense the reduction of stiffness than that in the armchair edge.

In addition, the maximum displacement in the entire graphene under shear stress in the zigzag edge is larger than that in the armchair edge as shown in Figures 11.9 and 11.10. Compared with the displacement results of graphene with center-concentrated vacancy defects, the influence of periodic regular vacancy defects in graphene is slighter. The change of displacement results in graphene with regular periodic vacancy defects loaded in the zigzag edge compared with that of pristine graphene is not as manifest as that in graphene with center-concentrated vacancy defects. Therefore, the equivalent Young's modulus is a more comprehensive and reliable

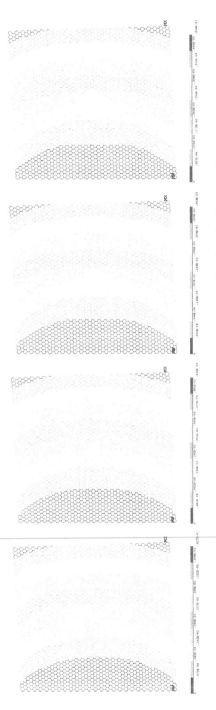

FIGURE 11.5 The displacement of graphene with center-concentrated vacancy defects under shear in the armchair edge.

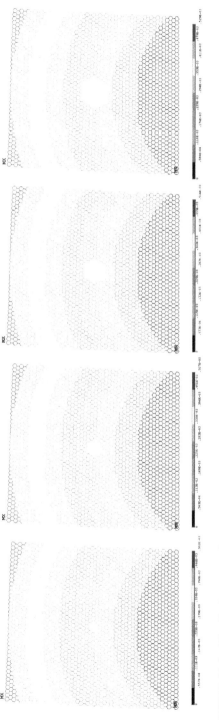

FIGURE 11.6 The displacement of graphene with center-concentrated vacancy defects under shear in the zigzag edge.

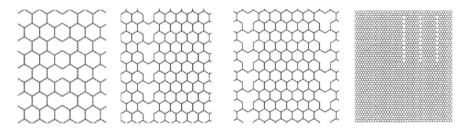

FIGURE 11.7 Periodic dispersion vacancy defects.

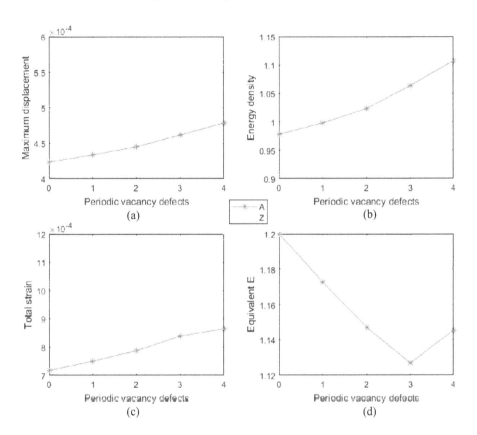

FIGURE 11.8 (a–d) The results of graphene with regular periodic vacancy defects (the units for a, b, and d are m, J/m³, and TPa, respectively).

factor to detect the impacts of vacancy defects than the parameters corresponding to local influence.

11.3.2 RANDOMLY DISTRIBUTED VACANCY DEFECTS

The negative effects of vacancy defects are not the unique impacts on the mechanical properties of graphene. The experimental measurements provide evidence of

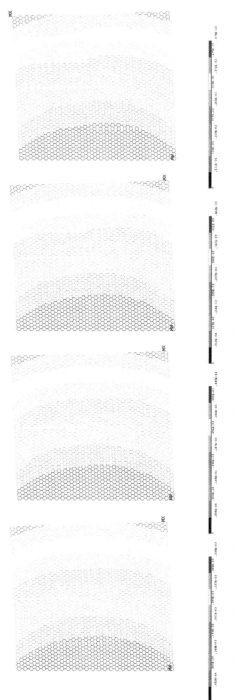

FIGURE 11.9 The displacement of graphene with regular periodic vacancy defects under shear in the armchair edge.

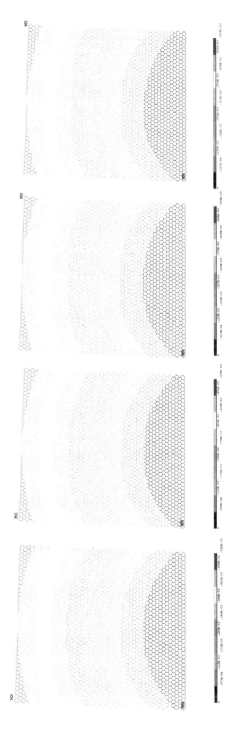

FIGURE 11.10 The displacement of graphene with regular periodic vacancy defects under shear in the zigzag edge.

strengthening effects by the inclusion of a low density of vacancy defects in graphene [13]. To confirm the possibility of strengthening effects by the vacancy defects in graphene, the vacancy defects are supposed to be randomly distributed and propagated in the entire graphene. By combing the finite element model with Monte Carlo simulation (MC-FEM), the vacancy defects are dispersed in the entire pristine graphene as shown in Figure 11.11.

In Figure 11.12, the results of graphene with randomly distributed vacancy defects are presented. The MC-FEM is repeated hundreds of times to stimulate the possible situations of vacancy defect dispersion. The maximum and minimum values of the equivalent Young's modulus, energy density, and total strain are captured in the computational procedure. The mean values of the mentioned parameters are also computed and compared in Figure 11.12.

For graphene under shear stress in the armchair edge, the equivalent Young's modulus has a maximum value at 1.240 TPa and a minimum value at 1.143 TPa, when the percentage of vacancy defects is 0.2%. If the amount of vacancy defects increases, the maximum values of the equivalent Young's modulus are 1.203, 1.214, and 1.191 TPa, for 0.5%, 0.8%, and 1% of vacancy defects, respectively. Furthermore, when the shear stress is loaded in the zigzag edge, the maximum values of the equivalent Young's modulus are 1.202, 1.195, 1.180, and 1.178 TPa, for 0.2%, 0.5%, 0.8%, and 1% of vacancy defects, respectively. For the same number of vacancy defects, graphene under shear stress in the armchair edge has the larger maximum equivalent Young's modulus. Moreover, the graphene under shear stress in the armchair edge has a larger probability to reach greater values of equivalent Young's modulus than that in the zigzag edge. The probabilities of stiffness strengthening effects in graphene under shear stress in the armchair edge are 2.2%, 0.4%, and 0.2%, for 0.2%, 0.5%, and 0.8% of vacancy defects, respectively. Besides, the probability of Young's modulus improvement is 1.8% in graphene under shear stress in the zigzag edge when the percentage of vacancy defects is as low as 0.2%. When the amount of vacancy defects is over 0.8% in graphene under shear stress in the armchair edge, the possibility of Young's modulus improvement is zero. In contrast, the enhancement effects of vacancy defects in the equivalent Young's modulus of graphene under uniaxial tension in the reported literature [28] are more evident than that under shear stress.

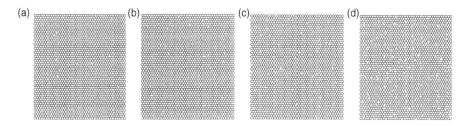

FIGURE 11.11   Randomly distributed vacancy defects (the percentages of vacancy defects in a–d are 0.2%, 0.5%, 1%, and 3%, respectively).

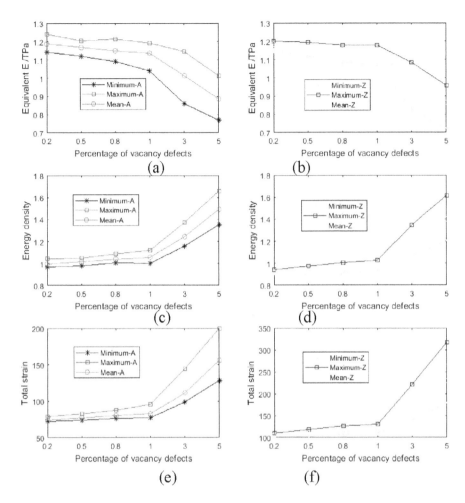

FIGURE 11.12 Results of graphene with randomly distributed vacancy defects (a, c, and e are for graphene under shear stress in the armchair edge; b, d, and f are for those in the zigzag edge).

The enhancement effects of vacancy defects in Young's modulus of graphene are an important impact on the research and application of graphene. However, the stiffness strengthening effects in vacancy-defected graphene under shear stress are not as evident as that under uniaxial tension. Not only the maximum equivalent Young's modulus for graphene with the same amount of vacancy defects, but also the probability of the enhancement effects is not comparable to the results in graphene under uniaxial tension.

For energy density and total strain, the maximum, minimum, and mean results are all magnified according to the increase of the number of vacancy defects under both boundary conditions. When the percentage of vacancy defects reaches 1% in Figure 11.12c–f, the results of energy density and total strain both have abrupt growth in graphene under two different boundary conditions. The randomly distributed vacancy defects contribute to the deviation in the equivalent Young's modulus, total strain, and energy density in the specified intervals. However, when the number of vacancy defects exceeds a certain value, namely 1% in this study, the influence of the amount of vacancy defects is drastic in equivalent Young's modulus, energy density, and total strain. Besides, owing to the randomly distributed location and the augment in the number of vacancy defects, the variance in the results of MC-FEM is enlarged.

To be more exact, the probability density distributions of graphene under two boundary conditions are illustrated in Figures 11.13 and 11.14. With the increase of the number of vacancy defects, Young's modulus varies in a wider interval in graphene under two different boundary conditions. Furthermore, the parts marked by circles in the probability density distribution in Figure 11.13 are smaller than those in Figure 11.14. This also proves that the probability of strengthening effects by vacancy defects in graphene under shear stress in the armchair edge is bigger than that in the zigzag edge but much less evident than that in graphene under uniaxial tension. In addition, the Gaussian probability distribution is an appropriate fitting result for the histogram of the equivalent Young's modulus, except when the amount of vacancy defects is as small as 0.2%, the histogram of the equivalent Young's modulus is more approximated to the confined peak, as shown in Figures 11.13 and 11.14.

For the displacement result analysis, one possible status of randomly distributed vacancy defects in graphene is performed with different certain amounts under two boundary conditions in Figures 11.15 and 11.16. When the amount of randomly distributed vacancy defects is minor, the influence of vacancy defects is not evident compared with that of pristine graphene for both boundary conditions. The independent displacement of each bond in graphene is not comparable to evaluate the impacts of vacancy defects in graphene under shear stress. The equivalent Young's

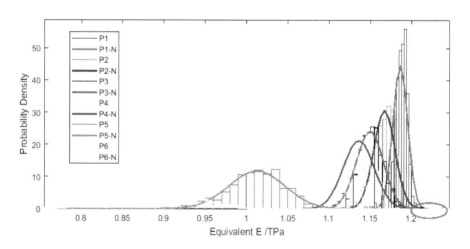

FIGURE 11.13 Probability density distribution of vacancy-defected graphene under shear stress in the armchair edge (P1–P6 are for 0.2%, 0.5%, 0.8%, 1%, 3%, and 5% of vacancy defects, respectively).

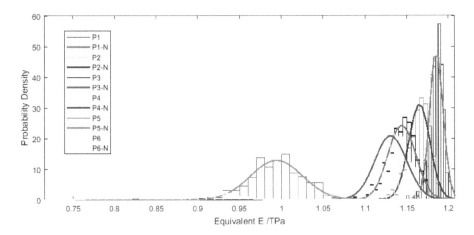

FIGURE 11.14 Probability density distribution of vacancy-defected graphene under shear stress in the zigzag edge (P1–P6 are for 0.2%, 0.5%, 0.8%, 1%, 3%, and 5% of vacancy defects, respectively).

modulus derived from energy density is feasible to predict the influences of vacancy defects in graphene.

The displacement in this study is focused on the in-plane deformation in graphene to discuss the influence of vacancy defects in the in-plane strength. For the out-plane deformation, Jalaei et al. [29] discussed the dynamic stability of an embedded orthotropic single-layer graphene sheet under the periodic excitation compressive load. The Kelvin–Voigt model is used to simulate the viscoelastic properties of graphene. The Ritz and Bolotin methods are conjunct to solve the governing equations of motion based on the energy method and Hamilton's principle for graphene. Besides, the stability for the graphene subjected to in-plane deformation is discussed. Baimova et al. [30] reported the stability range for a flat defect-free graphene sheet by neglecting the thermal vibrations and the effects of the boundary conditions. The instability is analyzed with respect to the phonon vibration modes within the first Brillouin zone. Bonilla and Carpio [31] developed the theory of defect dynamics in planar graphene. The periodic discrete elasticity is used to describe the defects in graphene as the cores or groups of dislocation. Moreover, Kuzubov et al. [32] discussed the dependences of stability and mobility of mono- and bi-vacancies on the degree of lattice deformation and temperature based on the density functional theory. Fang et al. [33] analyzed the wrinkling behavior of the defective graphene sheets under shear stress at different temperatures. The shear modulus and shear stress induced in the defective graphene sheet become larger when the temperature increases. Therefore, this work is meaningful to the parallel studies of vacancy defect impacts in graphene. More exploration on the vacancy defects influence on the stability of graphene will be performed in further work.

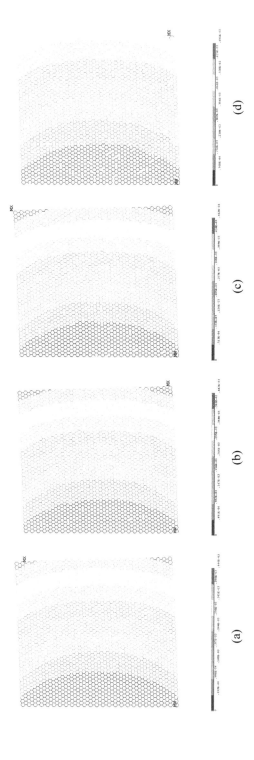

FIGURE 11.15   The displacement of graphene with randomly distributed vacancy defects under shear stress in the armchair edge (the percentages of vacancy defects in a–d are 0.2%, 0.5%, 1%, and 3%, respectively).

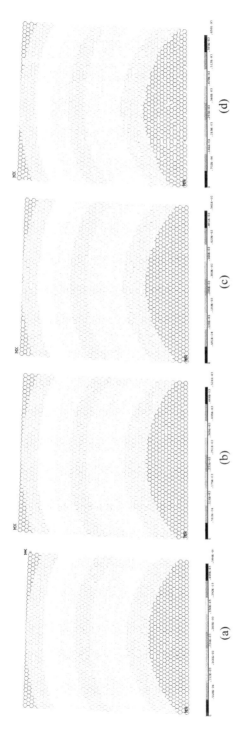

FIGURE 11.16 The displacement of graphene with randomly distributed vacancy defects under shear stress in the zigzag edge (the percentages of vacancy defects in a–d are 0.2%, 0.5%, 1%, and 3%, respectively).

11.4 CONCLUSION

This chapter predicts the influence of vacancy defects in graphene under shear stress. The location distribution and amounts of vacancy defects in the entire graphene are discussed and compared. From the results, the key points can be concluded as:

1. The center-concentrated vacancy defects evidently deteriorate the stiffness of graphene, and the equivalent Young's modulus is sharply cut down when the sizes of the center-concentrated vacancy defects become larger.
2. The impacts of vacancy defects are sensitive to the boundary condition. Graphene under shear stress in the zigzag edge is more robust to defense the reduction of Young's modulus caused by center-concentrated and periodic regular vacancy defects.
3. By MC-FEM, the probability of stiffness strengthening effects of vacancy defects in graphene is capable to be computed. The graphene under shear stress in the armchair edge has a larger probability to reach a larger equivalent Young's modulus than that in the zigzag edge.
4. The probability of strengthening effects in stiffness by vacancy defects in graphene under shear stress is small, as 2.2% and 1.8% under two boundary conditions, which is much less evident than that in graphene under uniaxial tension. Besides, when the amount of vacancy defects is amplified with certain ratios, the possibility of Young's modulus improvement disappears.

The equivalent Young's modulus is an appropriate and comprehensive factor to evaluate the influence of vacancy defects in graphene.

REFERENCES

1. Lee, C., Wei, X., Kysar, J.W., Hone, J. Measurement of the elastic properties and intrinsic strength of monolayer graphene. *Science*, **2008**, 3215887, 385–388.
2. Yanovsky, Y.G., Nikitina, E.A., Karnet, Y.N., et al. Quantum mechanics study of the mechanism of deformation and fracture of graphene. *Physical Mesomechanics*, **2009**, 125(6), 254–262.
3. Ni, Z., Bu, H., Zou, M., et al. Anisotropic mechanical properties of graphene sheets from molecular dynamics. *Physica B: Condensed Matter*, **2010**, 405(5), 1301–1306.
4. Tsai, J.L., Tu, J.F. Characterizing mechanical properties of graphite using molecular dynamics simulation. *Materials & Design*, **2010**, 311, 194–199.
5. Ansari, R., Ajori, S., Motevalli, B. Mechanical properties of defective single-layered graphene sheets via molecular dynamics simulation. *Superlattices and Microstructures*, **2012**, 512, 274–289.
6. Zhang, Y.Y., Wang, C.M., Cheng, Y., et al. 2011. Mechanical properties of bilayer graphene sheets coupled by sp3 bonding. *Carbon*, 4913, 4511–4517.
7. Javvaji, B., Budarapu, P.R., Sutrakar, V.K., et al. Mechanical properties of Graphene: Molecular dynamics simulations correlated to continuum based scaling laws. *Computational Materials Science*, **2016**, 125, 319–327.
8. Reddy, C.D., Rajendran, S., Liew, K.M. Equilibrium configuration and elastic properties of finite graphene. *Nanotechnology*, **2006**, 173, 864.

9. Georgantzinos, S.K., Giannopoulos, G.I., Anifantis, N.K. Numerical investigation of elastic mechanical properties of graphene structures. *Materials & Design*, **2010**, 3110, 4646–4654.

10. Scarpa, F., Adhikari, S., Phani, A.S. Effective elastic mechanical properties of single layer graphene sheets. *Nanotechnology*, **2009**, 206, 065709.

11. Shokrieh, M.M., Rafiee, R. Prediction of Young's modulus of graphene sheets and carbon nanotubes using nanoscale continuum mechanics approach. *Materials & Design*, **2010**, 312, 790–795.

12. Srivastava, A., Galande, C., Ci, L., et al. Novel liquid precursor-based facile synthesis of large-area continuous, single, and few-layer graphene films. *Chemistry of Materials*, **2010**, 2211, 3457–3461.

13. Dreyer, D.R., Park, S., Bielawski, C.W., Ruoff, R.S. The chemistry of graphene oxide. *Chemical Society Reviews*, **2010**, 391, 228–240.

14. Geim, A.K. Nobel lecture: Random walk to graphene. *Reviews of Modern Physics*, **2011**, 833, 851.

15. Zandiatashbar, A., Lee, G.H., An, S.J., et al. Effect of defects on the intrinsic strength and stiffness of graphene. *Nature Communications*, **2014**, 5, 3186.

16. Eckmann, A., Felten, A., Mishchenko, A., et al. Probing the nature of defects in graphene by Raman spectroscopy. *Nano Letters*, **2012**, 128, 3925–3930.

17. Ferrari, A.C., Meyer, J.C., Scardaci, V., et al. Raman spectrum of graphene and graphene layers. *Physical Review Letters*, **2006**, 9718, 187401.

18. Qin, H., Sun, Y., Liu, J.Z., et al. Mechanical properties of wrinkled graphene generated by topological defects. *Carbon*, **2016**, 108, 204–214.

19. Min, K., Aluru, N.R. Mechanical properties of graphene under shear deformation. *Applied Physics Letters*, **2011**, 981, 013113.

20. Papageorgiou, D.G., Kinloch, I.A., Young, R.J. Mechanical properties of graphene and graphene-based nanocomposites. *Progress in Materials Science*, **2017**, 90, 75–127.

21. Lópezpolín, G., Gómeznavarro, C., Parente, V., et al. Increasing the elastic modulus of graphene by controlled defect creation. *Nature Physics*, **2015**, 111, 26–31.

22. Belytschko, T., Xiao, S.P., Schatz, G.C., et al. Atomistic simulations of nanotube fracture. *Physical Review B Condensed Matter*, **2002**, 6523, 121–121.

23. Georgantzinos, S.K., Katsareas, D.E., Anifantis, N.K. Graphene characterization: A fully non-linear spring-based finite element prediction. *Physica E: Low-dimensional Systems and Nanostructures*, **2011**, 4310, 1833–1839.

24. Michele, M., Marco, R. Prediction of Young's modulus of single wall carbon nanotubes by molecular-mechanics based finite element modelling. *Composites Science and Technology*, **2006**, 6611, 1597–1605.

25. Chu, L., Shi, J., Souza de Cursi, E. Vibration analysis of vacancy defected graphene sheets by Monte Carlo based finite element method. *Nanomaterials*, **2018**, 87, 489

26. Fedorov, A.S., Popov, Z.I., Fedorov, D.A., et al. DFT investigation of the influence of ordered vacancies on elastic and magnetic properties of graphene and graphene-like SiC and BN structures. *Physica Status Solidi*, **2012**, 24912, 2549–2552.

27. Jing, N., Xue, Q., Ling, C., et al. Effect of defects on Young's modulus of graphene sheets: A molecular dynamics simulation. *RSC Advances*, **2012**, 224, 9124–9129.

28. Chu, L., Shi, J., Sun, L., Eduardo, S.D.C. The possibility of young's modulus improvement in graphene by random vacancy defects under uniaxial tension. *Materials Research Express*, **2019**, 6, 025007.

29. Jalaei, M.H., Arani, A.G., Tourang, H. On the dynamic stability of viscoelastic graphene sheets. *International Journal of Engineering Science*, **2018**, 132, 16–29.

30. Dmitriev, S.V., Baimova, Y.A., Savin, A.V., Kivshar, Y.S. Stability range for a flat graphene sheet subjected to in-plane deformation. *JETP Letters*, **2011**, 93(10) 571.

31. Bonilla, L.L., Carpio, A. Theory of defect dynamics in graphene: Defect groupings and their stability. *Continuum Mechanics and Thermodynamics*, **2011**, 23(4), 337–346.

32. Kuzubov, A.A., Anan'eva, Y.E., Fedorov, A.S., Tomilin, F.N., Krasnov, P.O. Quantum-chemical calculations on the stability and mobility of vacancies in graphene. *Russian Journal of Physical Chemistry A*, **2012**, 86(7), 1088–1090.

33. Fang, T.H., Chang, W.J., Lin, K.P., Shen, S.T. Stability and wrinkling of defective graphene sheets under shear deformation. *Current Applied Physics*, **2014**, 14(4), 533–537.

12 Strengthening Possibility by Random Vacancy Defects

12.1 INTRODUCTION

Graphene has two-dimensional (2D) honeycomb lattices of covalently bonded carbon atoms. An extraordinary intrinsic in-plane strength is observed in experiments and predicted in analytical theories. However, the reported Young's modulus of graphene is oscillated in a large interval and has not been settled as a certain value [1–11].

The deviation and fluctuation in the analytical and measured results of Young's modulus of graphene are mainly caused by the intrinsic and inescapable uncertainty or defects in the microstructure. The investigation of the influence of defects in graphene is an essential and challenging issue. On the one hand, the uncertainty and non-homogeneous properties are complicated due to different reasons, like uneven stress at boundaries, weakness of specific bonds, and Stone–Wales defects [12]. On the other hand, defects are unpredictable and randomly distributed in the entire graphene lattice. It is hard to control the size and placement of defects in the formation process of chemical vapor deposition [13], chemical reduction of graphene oxides [14], and mechanical exfoliation [15].

It is crucial to have more attention and investigation on graphene defects [16–20]. The uncertainty and complicity in the microstructure of graphene lead to the appearance of contradictory results or phenomena [21,22]. It is meaningful to confirm the possibility of stiffness improvement by a small amount of vacancy defect dispersion. The reasonable design of vacancy location distribution provides a promising technological relevance and guidance for the enhancement of mechanical properties.

In the present study, the possibility of Young's modulus improvement by the random vacancy defect dispersion in graphene is confirmed and discussed. The continuum mechanics approach is applied by performing the finite element model. The equivalent Young's modulus is computed from the elastic strain energy. The probability of strengthening effects by randomly distributed vacancy defects in graphene is presented and discussed. Besides, the maximum Young's modulus of vacancy-defected graphene is provided in this study, which is an important criterion for the optimization of mechanical properties in the design and manufacturing process. Finally, a summary is presented in the conclusion section.

12.2 MATERIALS AND METHODS

The modified Morse potential [23] can be used as in Chapter 11. With the spring finite element, the bond stretching and angle variations can be successfully

DOI: 10.1201/9781003226628-14

simulated [24,25]. The finite element model of pristine graphene is presented in Figure 12.1, which is based on the previous work of this study [26,27]. For each node, there are six degrees of freedom, namely displacement and rotation in the X, Y, and Z axes. The uniaxial tension is loaded in the armchair edge and zigzag edge, respectively. The contour results of vector sum displacement are also depicted in Figure 12.1 for pristine graphene. To discuss the effects of vacancy defects in the equivalent Young's modulus, the vacancy defects are randomly distributed in pristine graphene.

Monte Carlo simulation is used to propagate the vacancy defects in pristine graphene. The relative references demonstrating the bond-breaking defects possibly not only appear in the original production process but also play important roles in the fracture of graphene [28–30]. The amount of vacancy defects is counted by the C–C (carbon–carbon) bonds. The atom vacancy defects are special cases since one atom is related to three nearby C–C bonds. C–C bonds consist of the lattice structure of graphene with the capability to defend against deformation under external forces. The atomic vacancy defects are special cases and included in the bond-breaking model. The randomly distributed vacancy defects in pristine graphene are chosen and removed by Monte Carlo simulation. The formation of vacancy defects in pristine graphene changes the original regular honeycomb lattice graphene and influences the mechanical properties of graphene. By combing the finite element model with Monte Carlo simulation, the vacancy defects are created in the entire pristine graphene as shown in Figure 12.2.

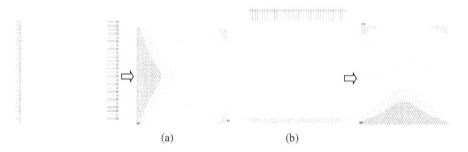

(a)   (b)

FIGURE 12.1   Finite element model of pristine graphene under uniaxial tension (a is for tension in the armchair edge and b is for tension in the zigzag edge).

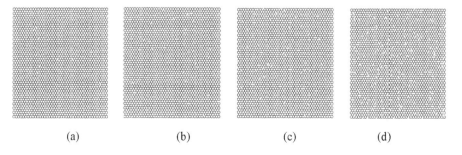

(a)   (b)   (c)   (d)

FIGURE 12.2   Random distributed vacancy defects (the percentages of vacancy defects in a–d are 0.2%, 0.5%, 1%, and 3%, respectively).

To observe the deformation of vacancy-defected graphene under uniaxial tension, the contour results of displacement vector sum for cases in Figure 12.2 are presented in Figures 12.3 and 12.4. By comparison with pristine graphene, vacancy-defected graphene with tension in the armchair edge has a more apparent difference than that in the zigzag edge. Besides, according to the increase of the vacancy defects, the influence of vacancy defects in graphene with tension in the armchair edge becomes larger. The deformation of vacancy-defected graphene is not only sensitive to the boundary condition (loading position) but also affected by the number of vacancy defects. However, the displacement computation for independent nodes in graphene lattice is not an appropriate method to comprehensively detect the influence of vacancy defects. Instead, the strain energy under tensile deformation in vacancy-defected graphene is calculated and captured as a feasible index to obtain the equivalent Young's modulus.

12.3 RESULTS AND DISCUSSION

Since the vacancy defects are randomly distributed in graphene, the deterministic finite element model is not sufficient to compute all possible situations of vacancy-defected graphene. The Monte Carlo-based finite element model is repeated hundreds of times to realize the stochastic dispersion of vacancy defects in pristine graphene. The maximum and minimum values of the equivalent Young's modulus, energy density, and total strain are computed in the sampling space.

Table 12.1 lists the statistical results of the equivalent Young's modulus in graphene with tension in the armchair and zigzag edges. For graphene with uniaxial tension in the armchair edge, the mean and minimum values of the equivalent Young's modulus both decrease with the growth of the amount of randomly distributed vacancy defects. However, the maximum values of the equivalent Young's modulus appear to have an amplified tendency when the number of vacancy defects is in a certain range as shown in Figure 12.5a. The possibility of Young's modulus improvement is confirmed in this situation. Moreover, the largest Young's modulus is obtained when the percentage of vacancy defects is 1%. For energy density and total strain, the maximum, minimum, and mean values are all magnified according to the increase in the number of vacancy defects. When the percentage of vacancy defects reaches 3% in Figure 12.5c and e, the maximum values of energy density and total strain both have abrupt growth.

For graphene with uniaxial tension in the zigzag edge, the largest Young's modulus is achieved when the percentage of vacancy defects is 0.5% as shown in Figure 12.5b. The minimum value of Young's modulus of graphene with 5% vacancy defects is smaller than that in graphene with tension in the armchair edge. In addition, for the energy density and total strain, the maximum values sharply increase when the percentage of vacancy defects reaches 3% as shown in Figure 12.5d and f. More complicatedly, the maximum energy density of Young's modulus has an obvious fluctuation when the percentage of vacancy defects is 0.5%.

The vacancy defects in graphene change the mechanical structure of the honeycomb lattice. On the one hand, the strain energy in graphene can be absorbed and clustered in the local places of vacancy defects. When the stain energy is partly reserved

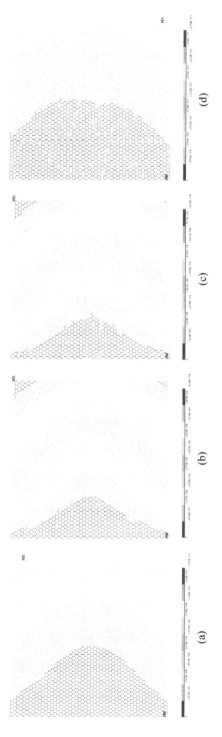

FIGURE 12.3 The displacement of graphene with randomly distributed vacancy defects under tension in the armchair edge (the percentages of vacancy defects in a–d are 0.2%, 0.5%, 1%, and 3%, respectively).

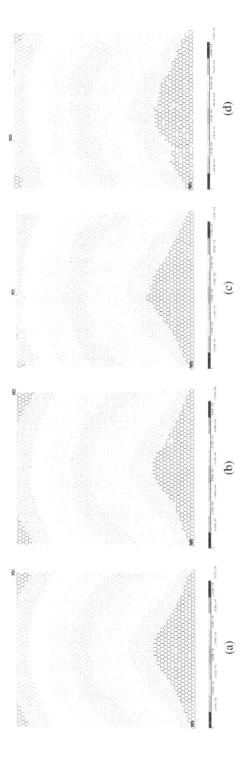

FIGURE 12.4 The displacement of graphene with randomly distributed vacancy defects under tension in the zigzag edge (the percentages of vacancy defects in a–d are 0.2%, 0.5%, 1%, and 3%, respectively).

TABLE 12.1

Statistic Results of the Equivalent Young's Modulus in Graphene under Uniaxial Tension

	Percentage	Min/TPa	Max/TPa	Mean/TPa	Variance	Probability
A	0.1%	1.1748	1.2525	1.1959	2.0499e-4	12.40%
	0.2%	1.1575	1.2849	1.1917	3.2015e-4	12.20%
	0.5%	1.1298	1.3889	1.1800	8.5979e-4	20.20%
	1%	1.0870	1.4319	1.1596	0.0016	12.80%
	3%	0.9230	1.3169	1.0582	0.0038	3.0%
	5%	0.7812	1.1315	0.9331	0.0037	0
Z	0.1%	1.1712	1.2241	1.1922	3.3712e-5	9.20%
	0.2%	1.1612	1.2335	1.1855	7.4665e-5	5.20%
	0.5%	1.1243	1.4510	1.1654	3.7885e-4	0.80%
	1%	1.0748	1.3237	1.1320	6.2077e-4	1.00%
	3%	0.6305	1.2648	0.9986	0.0022	0.60%
	5%	0.5406	1.1942	0.8667	0.0037	0

in the place of vacancy defects, the deformation energy in the rest part of graphene is reduced. The displacement in places except for the positions of vacancy defects becomes smaller. The enhancement effects can be observed when the vacancy defects are reasonably distributed. On the other hand, the local position of vacancy defects is more delicate to have damages than other parts of graphene. It is necessary to ensure safety in the local place of vacancy defects when amplifying the enhancement effects.

The equivalent Young's modulus is the comprehensive result of energy and total strain of the entire graphene. The fluctuations in the equivalent Young's modulus marked by red circles in Figure 12.5 are the trade-off between mechanical energy and total strain. The changes in mechanical energy and total strain in graphene are due to the introduction of vacancy defects. The structural integrity of pristine graphene is destroyed by the randomly distributed vacancy defects. Among tremendous samples in the Monte Carlo-based finite element model, the maximum value of the equivalent Young's modulus is computed and recorded. In the first stage, when the amount of vacancy defects is tiny, the influence on the mechanical properties of graphene is not apparent under two boundary conditions. In the second stage, with the amplification of the number of vacancy defects, both the energy and total strain in graphene have increment. However, the degree of increase is not consistent in energy and total strain. The increasing percentage of energy is larger than that of total strain in the second stage. Thus, the peaks of the equivalent Young's modulus appear in graphene under both boundary conditions. In the third stage, the percentage of augmentation in mechanical energy caused by vacancy defects cannot balance the total strain. The maximum equivalent Young's modulus decreases with the increase of vacancy defects. The enhancement effects with the non-linear relationship with the defect amount are also observed in Ref. [22]. The results of the Monte Carlo simulation have a good agreement with experimental data.

To be more exact, the probability density distribution of graphene under two boundary conditions is illustrated in Figure 12.6. No matter what uniaxial tension

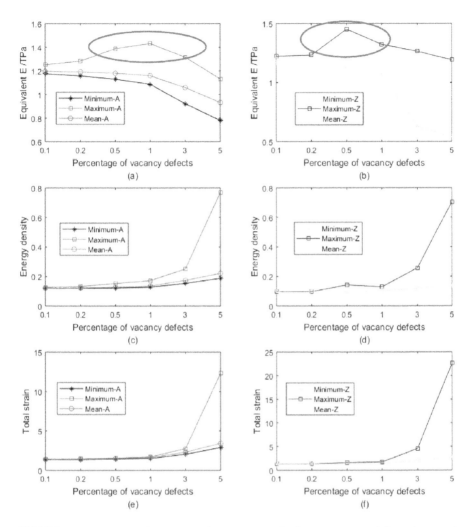

FIGURE 12.5 Results of graphene with randomly distributed vacancy defects (a, c, and e are for graphene with uniaxial tension in the armchair edge; b, d, and f are for that in the zigzag edge).

is in the armchair or the zigzag edge, with the increase of the number of vacancy defects, Young's modulus varies in a wider interval. The variance results in Table 12.1 also prove this point. Besides, the probability of enhancement effects by vacancy defects in graphene under uniaxial tension in the armchair edge is bigger than that in the zigzag edge. The results of probability density distribution in random vacancy-defected graphene with tension in the armchair edge appear on the right side of that with tension in the zigzag edge. Furthermore, the interval range of the equivalent Young's modulus moves to the right side at first with the augmentation of the number of vacancy defects. But when the number of vacancy defects continues to increase, the interval range of the equivalent Young's modulus returns back and transfers to the left side. Moreover, even the equivalent Young's modulus reaches a maximum

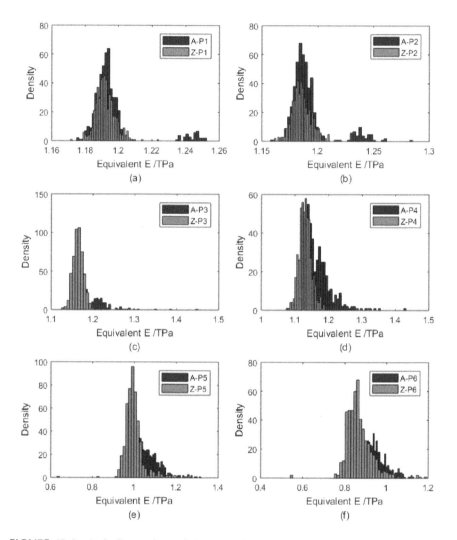

FIGURE 12.6 (a–f) Comparison of the probability density distribution of the equivalent Young's modulus under two boundary conditions (P1–P6 are for 0.1%, 0.2%, 0.5%, 1%, 3%, and 5% of vacancy defects, respectively).

value with the largest probability as shown in Figure 12.6.c; the second peaks of probability density distribution existing in Figures 12.6a and b are non-negligible. The information of the maximum equivalent Young's modulus and largest probability related to the amount of vacancy defects are useful references and guidance to the optimization process of mechanical property improvement for graphene. The second peak in the probability density distribution of the equivalent Young's modulus provides a reasonable explanation for the deviation and fluctuation of the tested results of Young's modulus in the experiments. When the number of vacancy defects in graphene is small, such as 0.1%–0.2%, the probability to have deviated results in another independent interval range is considerable.

To be more obvious, the cumulative probability of the equivalent Young's modulus is computed for graphene with different amounts of vacancy defects and presented in Figure 12.7. The maximum equivalent Young's modulus of vacancy-defected

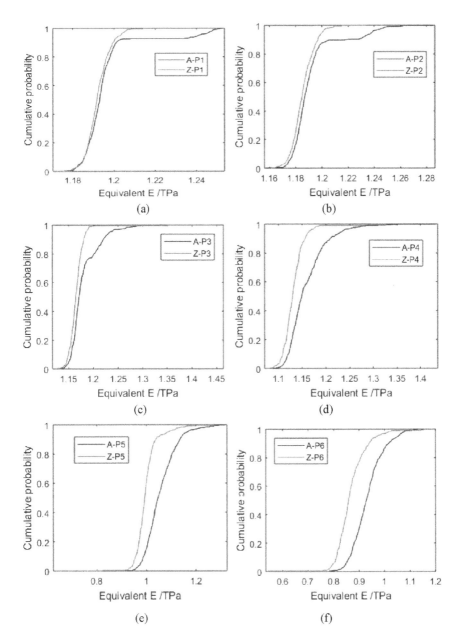

FIGURE 12.7   (a–f) Comparison of the cumulative probability of the equivalent Young's modulus under two boundary conditions (P1–P6 are for 0.1%, 0.2%, 0.5%, 1%, 3%, and 5% of vacancy defects, respectively).

graphene with tension in both the zigzag and armchair edges can be 1.4 TPa. Even though the maximum Young's modulus of graphene under two different boundary conditions are as close as that in Figure 12.8a, the probability of strengthening effects in graphene with tension in the armchair edge is absolutely higher than that in the zigzag edge as shown in Figures 12.7 and 12.8b. For graphene with tension in the armchair edge, the strengthening effects are more possible to occur when the percentage of vacancy defects is 0.5%, with the probability close to 20% in Figure 12.7c. The probability of strengthening effects is amplified with the increase of the number of vacancy defects at the beginning. When the probability reaches the peak as mentioned above, the probability of strengthening effects is cut down with the enlargement of the number of vacancy defects. For graphene with tension in the zigzag edge, the probability of strengthening effects has a monotonous descending trend. For both boundary conditions, the probability of strengthening effects is zero when the number of vacancy defects is 5%.

The differences in graphene with tension in the armchair and zigzag edges are important. The discrepancies in the probability of enhancement effects in graphene

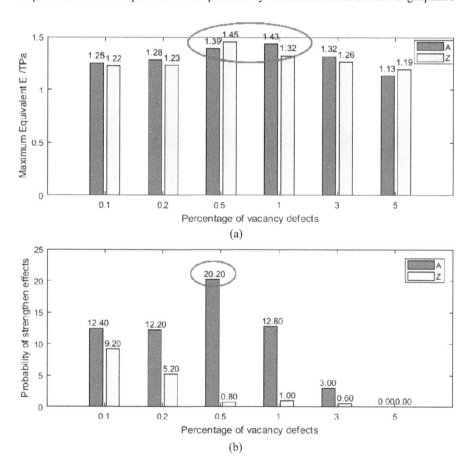

FIGURE 12.8   (a and b) Probability criterion results of vacancy-defected graphene.

are not only due to the orientation in the pristine microstructure of the honeycomb lattice but also caused by the boundary conditions. The loading location in geometrical characteristics and the total external uniaxial tension force is the key factor of boundary conditions to play roles in the mechanical results. However, the results of the maximum equivalent Young's modulus in graphene with tension in the armchair and zigzag edges are consistent. The feasibility of multiple objective optimizations in graphene by a reasonable vacancy defect design is well supported. It is possible to have satisfied enhancement effects by vacancy defects in graphene, no matter the uniaxial tension is loaded in the armchair or zigzag edges.

12.4 CONCLUSION

This work confirmed the possibility of Young's modulus improvement in graphene by random vacancy defects under uniaxial tension. The probability of enhancement effects and the number of vacancy defects in graphene are discussed and compared in this chapter. From the Monte Carlo-based finite element model, the probability of enhancement effects in stiffness by vacancy defects in graphene is provided. When the percentage of vacancy defects in graphene with uniaxial tension in the armchair edge is 0.5%, the probability of strengthening effects is as large as 20.5%, which is much greater than that in graphene with tension in the zigzag edge. Even though the maximum values of Young's modulus under both boundary conditions have the possibility to reach 1.4 TPa, the corresponding numbers of vacancy defects are different, which happen in 1% and 0.5% for tension in the armchair and zigzag edges, respectively. The work in this study provides an important criterion for the optimization of mechanical properties in the production and manufacturing process of graphene.

REFERENCES

1. Lee, C., Wei, X., Kysar, J.W., Hone, J. Measurement of the elastic properties and intrinsic strength of monolayer graphene. *Science*, **2008**, 3215887, 385–388.
2. Yanovsky, Y.G., Nikitina, E.A., Karnet, Y.N., et al. Quantum mechanics study of the mechanism of deformation and fracture of graphene *Physical Mesomechanics*, **2009**, 125(6), 254–262.
3. Ni, Z., Bu, H., Zou, M., et al. Anisotropic mechanical properties of graphene sheets from molecular dynamics. *Physica B: Condensed Matter*, **2010**, 405(5), 1301–1306.
4. Tsai, J.L., Tu, J.F. Characterizing mechanical properties of graphite using molecular dynamics simulation. *Materials & Design*, **2010**, 311, 194–199.
5. Ansari, R., Ajori, S., Motevalli, B. Mechanical properties of defective single-layered graphene sheets via molecular dynamics simulation. *Superlattices and Microstructures*, **2012**, 512, 274–289.
6. Zhang, Y.Y., Wang, C.M., Cheng, Y., et al. Mechanical properties of bilayer graphene sheets coupled by sp3 bonding. *Carbon*, **2011**, 4913, 4511–4517.
7. Javvaji, B., Budarapu, P.R., Sutrakar, V.K., et al. Mechanical properties of Graphene: Molecular dynamics simulations correlated to continuum based scaling laws. *Computational Materials Science*, **2016**, 125, 319–327.
8. Reddy, C.D., Rajendran, S., Liew, K.M. Equilibrium configuration and elastic properties of finite graphene. *Nanotechnology*, **2006**, 173, 864.

9. Georgantzinos, S.K., Giannopoulos, G.I., Anifantis, N.K. Numerical investigation of elastic mechanical properties of graphene structures. *Materials & Design*, **2010**, 3110, 4646–4654.

10. Scarpa, F., Adhikari, S., Phani, A.S. Effective elastic mechanical properties of single layer graphene sheets. *Nanotechnology*, **2009**, 206, 065709.

11. Shokrieh, M.M., Rafiee, R. Prediction of Young's modulus of graphene sheets and carbon nanotubes using nanoscale continuum mechanics approach. *Materials & Design*, **2010**, 312, 790–795.

12. Deng, S., Berry, V. Wrinkled, rippled and crumpled graphene: An overview of formation mechanism, electronic properties, and applications. *Materials Today*, **2016**, 19(4), 197–212.

13. Srivastava, A., Galande, C., Ci, L., et al. Novel liquid precursor-based facile synthesis of large-area continuous, single, and few-layer graphene films. *Chemistry of Materials*, **2010**, 2211, 3457–3461.

14. Dreyer, D.R., Park, S., Bielawski, C.W., Ruoff, R.S. The chemistry of graphene oxide. *Chem Soc Reviews*, **2010**, 391, 228–240.

15. Geim, A.K. Nobel lecture: Random walk to graphene. *Reviews of Modern Physics*, 2011, 833, 851.

16. Zandiatashbar, A., Lee, G.H., An, S.J., et al. Effect of defects on the intrinsic strength and stiffness of graphene. *Nature Communications*, **2014**, 5, 3186.

17. Eckmann, A., Felten, A., Mishchenko, A., et al. Probing the nature of defects in graphene by Raman spectroscopy. *Nano Letters*, **2012**, 128, 3925–3930.

18. Ferrari, A.C., Meyer, J.C., Scardaci, V., et al. Raman spectrum of graphene and graphene layers. *Physical Review Letters*, **2006**, 9718, 187401.

19. Qin, H., Sun, Y., Liu, J.Z., et al. Mechanical properties of wrinkled graphene generated by topological defects. *Carbon*, **2016**, 108, 204–214.

20. Min, K., Aluru, N.R. Mechanical properties of graphene under shear deformation. *Applied Physics Letters*, **2011**, 981, 013113.

21. Papageorgiou, D.G., Kinloch, I.A., Young, R.J. Mechanical properties of graphene and graphene-based nanocomposites. *Progress in Materials Science*, **2017**, 90, 75–127.

22. Lópezpolín, G., Gómeznavarro, C., Parente, V., et al. Increasing the elastic modulus of graphene by controlled defect creation. *Nature Physics*, **2015**, 111, 26–31.

23. Belytschko, T., Xiao, S.P., Schatz, G.C., et al. Atomistic simulations of nanotube fracture. *Physical Review B Condensed Matter*, **2002**, 6523, 121–121.

24. Georgantzinos, S.K., Katsareas, D.E., Anifantis, N.K. Graphene characterization: A fully non-linear spring-based finite element prediction. *Physica E: Low-dimensional Systems and Nanostructures*, **2011**, 4310, 1833–1839.

25. Michele, M., Marco, R. Prediction of Young's modulus of single wall carbon nanotubes by molecular-mechanics based finite element modelling. *Composites Science and Technology*, **2006**, 6611, 1597–1605.

26. Chu, L., Shi, J., Souza de Cursi, E. Vibration analysis of vacancy defected graphene sheets by Monte Carlo based finite element method. *Nanomaterials*, **2018**, 87, 489.

27. Chu, L., Shi, J., Ben, S. Buckling analysis of vacancy-defected graphene sheets by the Stochastic finite element method. *Materials*, **2018**, 11(9), 1545.

28. Geim, A.K. Graphene: Status and prospects. *Science*, **2009**, 324(5934), 1530–1534.

29. Grantab, R., Shenoy, V.B., Ruoff, R.S. Anomalous strength characteristics of tilt grain boundaries in graphene. *Science*, **2010**, 330(6006), 946–948.

30. Terdalkar, S.S., Huang, S., Yuan, H., Rencis, J.J., Zhu, T., Zhang, S. Nanoscale fracture in graphene. *Chemical Physics Letters*, **2010**, 494(4–6), 218–222.

Index

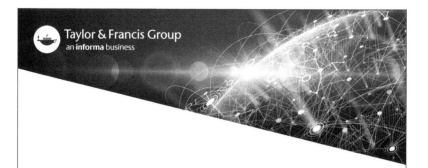

Taylor & Francis Group
an **informa** business

Taylor & Francis eBooks

www.taylorfrancis.com

A single destination for eBooks from Taylor & Francis
with increased functionality and an improved user
experience to meet the needs of our customers.

90,000+ eBooks of award-winning academic content in
Humanities, Social Science, Science, Technology, Engineering,
and Medical written by a global network of editors and authors.

TAYLOR & FRANCIS EBOOKS OFFERS:

A streamlined
experience for
our library
customers

A single point
of discovery
for all of our
eBook content

Improved
search and
discovery of
content at both
book and
chapter level

REQUEST A FREE TRIAL
support@taylorfrancis.com

 Routledge
Taylor & Francis Group

 CRC Press
Taylor & Francis Group

Milton Keynes UK
Ingram Content Group UK Ltd.
UKHW022040141024
449569UK00014B/669